Tanja Radic

Circadian system in Neurospora crassa wild-type isolates

AF062058

Tanja Radic

Circadian system in Neurospora crassa wild-type isolates

Genetics and physiology

Südwestdeutscher Verlag für Hochschulschriften

Impressum/Imprint (nur für Deutschland/only for Germany)
Bibliografische Information der Deutschen Nationalbibliothek: Die Deutsche Nationalbibliothek verzeichnet diese Publikation in der Deutschen Nationalbibliografie; detaillierte bibliografische Daten sind im Internet über http://dnb.d-nb.de abrufbar.
Alle in diesem Buch genannten Marken und Produktnamen unterliegen warenzeichen-, marken- oder patentrechtlichem Schutz bzw. sind Warenzeichen oder eingetragene Warenzeichen der jeweiligen Inhaber. Die Wiedergabe von Marken, Produktnamen, Gebrauchsnamen, Handelsnamen, Warenbezeichnungen u.s.w. in diesem Werk berechtigt auch ohne besondere Kennzeichnung nicht zu der Annahme, dass solche Namen im Sinne der Warenzeichen- und Markenschutzgesetzgebung als frei zu betrachten wären und daher von jedermann benutzt werden dürften.

Coverbild: www.ingimage.com

Verlag: Südwestdeutscher Verlag für Hochschulschriften GmbH & Co. KG
Heinrich-Böcking-Str. 6-8, 66121 Saarbrücken, Deutschland
Telefon +49 681 37 20 271-1, Telefax +49 681 37 20 271-0
Email: info@svh-verlag.de

Approved by: München, LMU, Diss, 2012

Herstellung in Deutschland (siehe letzte Seite)
ISBN: 978-3-8381-3419-2

Imprint (only for USA, GB)
Bibliographic information published by the Deutsche Nationalbibliothek: The Deutsche Nationalbibliothek lists this publication in the Deutsche Nationalbibliografie; detailed bibliographic data are available in the Internet at http://dnb.d-nb.de.
Any brand names and product names mentioned in this book are subject to trademark, brand or patent protection and are trademarks or registered trademarks of their respective holders. The use of brand names, product names, common names, trade names, product descriptions etc. even without a particular marking in this works is in no way to be construed to mean that such names may be regarded as unrestricted in respect of trademark and brand protection legislation and could thus be used by anyone.

Cover image: www.ingimage.com

Publisher: Südwestdeutscher Verlag für Hochschulschriften GmbH & Co. KG
Heinrich-Böcking-Str. 6-8, 66121 Saarbrücken, Germany
Phone +49 681 37 20 271-1, Fax +49 681 37 20 271-0
Email: info@svh-verlag.de

Printed in the U.S.A.
Printed in the U.K. by (see last page)
ISBN: 978-3-8381-3419-2

Copyright © 2012 by the author and Südwestdeutscher Verlag für Hochschulschriften GmbH & Co. KG and licensors
All rights reserved. Saarbrücken 2012

TABLE OF CONTENT

1. INTRODUCTION 3
   - Biological rhythms 3
     - Circadian rhythms 4
     - Molecular mechanism of circadian rhythm 6
     - Properties of circadian rhythms 7
     - Entrainment 9
     - Relevance of circadian rhythms 10
   - Photoperiodism 12
   - Models of daylength measurements 14
   - *Neurospora crassa* 16
     - Life cycle 19
     - Natural habitat 20
     - Evolution 21
     - Circadian clock 23
   - Aims of the study 28
2. MATERIALS AND METHODS 29
   - Strains 29
   - Strain maintenance 30
   - Molecular methods 31
     - DNA preparation 31
     - Amplification of DNA with PCR 31
     - DNA sequencing 33
       - Sequencing setup 33
       - Phylogenetic analysis 33
   - Physiological methods 34
     - Growth conditions 34
       - Light-dark cycles 34
       - Temperature cycles 35
     - Race tube assay 35
       - Race tube setup 35
       - Data analysis 36
   - Statistical analysis 37
3. RESULTS 39
   - Genetic analysis 39
     - Neutral markers 39
     - *White collar-1* gene 46
     - *Frequency* locus 51
     - *Frequency* promoter 52
     - Comparison between clock genes and clades phylogeny 55
   - Physiology 59
     - The clock of isolates in constant conditions 59
     - The clock in entrained conditions 62
       - Cluster analysis 62
       - Variability of the data 69
       - Comparison between light and temperature surface 70
   - Association between DNA sequences and circadian phenotypes 72
     - Free-running periods 72
     - Clock in entrained conditions 74
4. DISCUSSION 78
   - Grouping of the isolates 78
     - Geographical distribution of isolates across clades 78

    Phylogenetic analysis of clock genes: Comparison with clades 79
    Physiology 80
        Comparison between physiology and genetics 81
    Latitudinal cline 81
        Genetics 81
        Physiology 83
CONCLUSIONS 85
SUMMARY 86
ZUSAMMENFASSUNG 88
REFERENCES 90
APPENDIX 106
    Abbreviations 106
    Recipes 108

1. INTRODUCTION

Biological rhythms

Coping with the environment is a crucial attribute of many, if not all organisms. Environment fluctuates according to the time of day and year, giving temporal information to the organisms. They use this information to synchronize their biology in order to be prepared in advance to advantageous or disadvantageous seasons (Bradshaw and Holzapfel, 2007). As a consequence, organisms evolved a mechanism of biological rhythms. The biological rhythms are observed at cellular and molecular level (Rusak and Zucker, 1975) and confer correct timing within the environment. An organism has to adapt to environmental change for its survival. Therefore, biological rhythm synchronizes the cellular behaviour to external geophysical time cues and contributes a selective advantage to an organism by allowing it to optimize for environmental changes (Edery, 2000). The environmental cues that are perceived are called *zeitgebers*, from the German word meaning "time giver". The internal mechanism able to track environmental temporal processes is termed a biological clock and is responsible for maintaining rhythmic phenomenon in the absence of the environmental stimulus (Rusak and Zucker, 1975).

The length of light and darkness is directly linked to the daily earth rotation and the annual change of inclination of the earth's rotational axis in relation to the sun. These geophysical features create environmental time cues, which biological rhythms depend on. There are four different kinds of biological rhythms described by their phenomenology, depending principally on period length (Pittendrigh, 1993):

- Yearly or circannual rhythms of 365.25 days, caused by the rotation of the earth around the sun, for example gonad development in birds (Rowan, 1925).
- Lunar rhythms of about 28.5 days caused by the rotation of the moon around the earth, for example marine reproduction of polychaete *Typosyllis prolifera* (Franke, 1985).
- Daily or circadian rhythms of 24 hours, caused by the rotation of the earth around its own axis. In specific, rhythms shorter than 24 hours are called ultradian; rhythms longer than 24 hours are called infradian. An example of circadian rhythm is sleep-wake cycle in humans (Aschoff and Wever, 1980).
- Tidal rhythms of about 12.5 hours, generated by the gravitational pull of the moon, for example the crab activity on shoreline (Naylor, 1996).

All of these rhythms are very important because they affect the activity of organisms and thereby increasing their fitness.

Biological rhythms can be generated in two ways (Aschoff, 1960):

- Exogenous rhythms are directly driven by environmental or other external rhythms; e.g., perch-hopping of sparrows elicited by light (Binkley and Mosher, 1985).
- Endogenous rhythms are driven by an internal self-sustaining biological clock, rather than by an external process. These rhythms will be maintained even if the environmental cues are removed, e.g., core body temperature (Aschoff, 1982). The endogenous rhythms are the most interesting since they are provide timing within the environment and therefore give advantages to the organism.

Biological clocks have evolved because precisely timed rhythmic activities gain adaptive advantage. Clocks function to: (1) restrict the activity to species-specific times of the day (for example when males and females of the same species become fertile so the period of fertility coincides), (2) reduce intrinsic competitive disadvantage for the inferior competitor (for example prey species will be active at times when they think the predators are not), and (3) give temporal memory so an organism can be prepared for environmental changes (Paranjpe and Sharma, 2005).

Circadian rhythms

The term circadian rhythms, proposed by Franz Halberg in late 1950s, refers to biological rhythms, which have a period of about a day. In the absence of environmental cues (*zeitgebers*), it period is approximately 24 hours - giving the name circadian rhythms (Lat. *circa* = about; *dies* = day). This rhythm is generated endogenously, because it persists even when the organism is placed in constant conditions (e.g., continuous darkness) (Merrow *et al.*, 2005).

The first description of a circadian rhythm was from the French astronomer Jean Jacques d'Ortous de Mairan in the 1700s. He noticed a 24-hours pattern in the movement of the leaves of the plant *Mimosa pudica,* which continued even in the absence of external when the plant was put in a dark box (Figure 1.1; de Mairan, 1729). Also Darwin described in his book The Power of Movement in Plants (1880) how the leaves of mimosa close in the evening and open in the morning. However, 200 years later Erwin Bünning demonstrated that period length is heritable and gave the first evidence for the genetic basis of circadian rhythms in plants (Bünning, 1962). From plants the field of circadian rhythms was extended also for other organisms like bacteria, rodents, insects, birds, primates and also humans (Dunlap, 1999).

Figure 1.1: In 1729, de Mairan demonstrated the existence of circadian rhythms in *Mimosa pudica*. He was fascinated by the daily opening and closing of the leaves and performed an experiment where he put the plant in constant darkness and then observed the behaviour. He saw that the rhythmic opening and closing of the leaves continued even in the absence of sunlight. Despite his result, de Mairan hesitated to conclude that *Mimosa* had an internal clock and hypothesized that other factors, such as temperature and magnetic field, were responsible for the rhythmic behaviour and decided not to publish his results (Text source: Wikipedia). Figure downloaded on 15.05.2011 from http://cronobio.es/Historia/deMairan.htm.

Circadian rhythms regulate gene expression, metabolic processes, activity and reproduction in order to coordinate biological processes with exogenous environmental cycles (Sharma, 2003). Some of the most studied circadian processes are cell division in cyanobacteria (Dong *et al.*, 2010), melatonin levels in birds (Binkley *et al.*, 1977), activity/rest cycles in mammals (Huang *et al.*, 2011), conidiation in *Neurospora* (Bell-Pedersen *et al.*, 2005), hatching in *Drosophila* (Manjunatha *et al.*, 2008) and stomata opening and leaf movements in plants (Yakir *et al.*, 2007). A substantial percentage of the genome is circadianly regulated, for example around one third of the plant genome (Covington *et al.*, 2008) and a quarter of the *Neurospora* genome (Dong *et al.*, 2008).

The simplified model of a circadian system comprises three basic components (Figure 1.2; Roenneberg *et al.*, 1998):

- Input pathways: *zeitgebers* activate the receptors and they transfer this information to the oscillator. Light is one of the main environmental time cues for the circadian systems. Also non-photic *zeitgebers*, such as ambient temperature, food availability, physical activity, and social contact can activate circadian system.
- Central oscillator, which generates the circadian rhythms. Photic and non-photic *zeitgebers* act on different circadian pacemakers, suggesting that several oscillators drive the circadian program (Daan *et al.*, 1984).

- Output pathways represented by regulation of *clock-controlled genes* (*ccg*), which modulate many physiological properties.

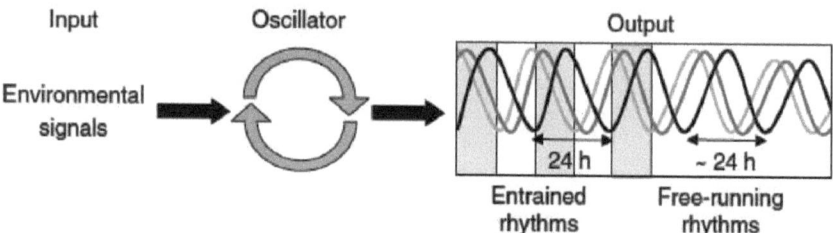

Figure 1.2: The three components of the basic circadian model: input pathways, central oscillator and output pathways. Taken from Yerushalmi and Green, 2009.

Circadian rhythms have been observed in almost all organisms from prokaryotic unicellular cyanobacteria to higher mammals (Yerushalmi and Green, 2009), including microorganisms such as algae, fungi, and protozoa (Sweeney, 1976). Given that circadian rhythms have been identified even in a single-cellular algae *Gonyaulax polyedra* (Hastings and Dunlap, 1986), suggests that they evolved early, when life consisted of single cells (Anders, 1982). It is believed that circadian rhythms of the earliest cells provided a mechanism for protection of DNA from ultraviolet radiation (Tauber *et al.*, 2004).

Molecular mechanism of circadian rhythm

At the molecular level, the core of the clocks consists of a transcription/translation oscillator (Bell-Pedersen *et al.*, 2005). There are two levels of regulation: negative feedback and post-transcriptional control. A network of positive and negative elements establishes the negative feedback loops generating the basic rhythmicity (Figure 1.3). In a simple view, every oscillator has both positive and negative elements that form the feedback loop. The positive elements of the loop activate the expression of the negative elements, while the negative elements feedback to block their own activation by the positive elements. These positive elements are transcriptional factors, which have PAS domain to form heterodimeric complexes and are sensitive to external stimuli (Crosthwaite *et al.*, 1997; Allada *et al.*, 1998; Rutila *et al.*, 1998; King *et al.*, 1997; Kay, 1997). Post-transcriptional control of the genes allows the clock to cycle with a 24 hours period and provides a way for the clock to be reset by *zeitgebers*. There are four different ways of how post-transcriptional control works: (1) control of RNA (for example different splicing; Cheng *et al.*, 1998), (2) translation control (Jackson *et al.*, 2010), (3) shuffling of the protein (for example from cytoplasm to the nucleus; Tataroglu and Schafmeier, 2010), and (4) protein degradation (Syed *et al.*, 2011). This is a very simplified way to describe the clock, a single central loop is unlikely: in reality a clock is composed of a network of feedback loops (Roenneberg and Merrow, 2003).

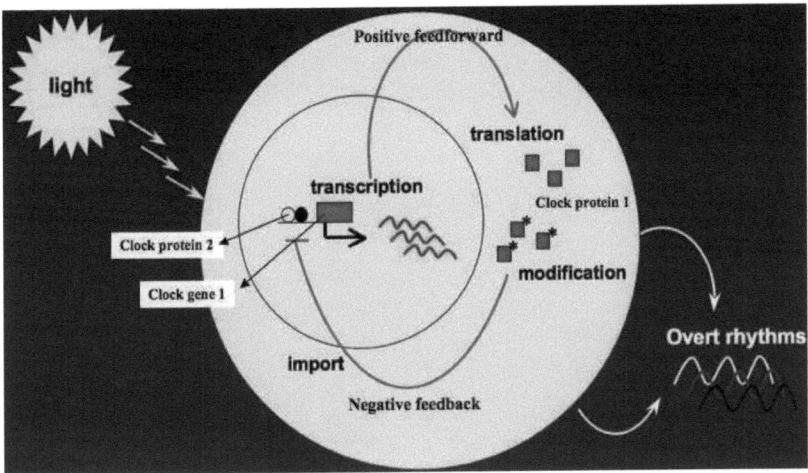

Figure 1.3: Generic feedback loop model. Clock gene 1 is transcribed into RNA and translated into protein (red squares). Clock protein 2 positively regulates transcription of clock gene 1. Clock protein 1 negatively regulates its own transcription by interfering with the positive effect of clock protein 2. Figure downloaded on 15.06.2011 and redrawn from http://rumba.biosci.ohio-state.edu/rumba_best_somers_2008.html.

However, this model has recently been challenged by observation of the simplest known circadian clock in prokaryotic cyanobacteria *Synecohococcus elongatus*, which can be reconstituted *in vitro* with just the three proteins of their central oscillator and ATP (Nakajiama *et al.*, 2005). Moreover, circadian rhythms are found in human red blood cells, which have no nucleus and therefore cannot perform transcription, indicating that transcription is not required for circadian oscillators in humans (O'Neill and Reddy, 2011).

The organization of the circadian system differs between organisms. In plants, almost every cell has its own autonomous clock (Thain *et al.*, 2002). The mammals have peripheral oscillators in tissues (for example, in the liver) that are synchronized by a central oscillator, which controls circadian physiology. This central oscillator, localized in the nervous systems, is a small group of neurons in the hypothalamus, called suprachiasmatic nucleus (SCN) (Dibner *et al.*, 2010).

Properties of circadian rhythms

Circadian oscillators can be described with three fundamental parameters (Figure 1.4): cycle length (T), which is the length of the external oscillator (ranges from one certain point of the first oscillation to the same point in the next cycle); period (τ) is the time required to progress through one circadian cycle (one cycle ranges from one defined clock marker to the next one), and phase (φ) is the time between external and internal oscillator ($\Delta\varphi = \tau - T$).

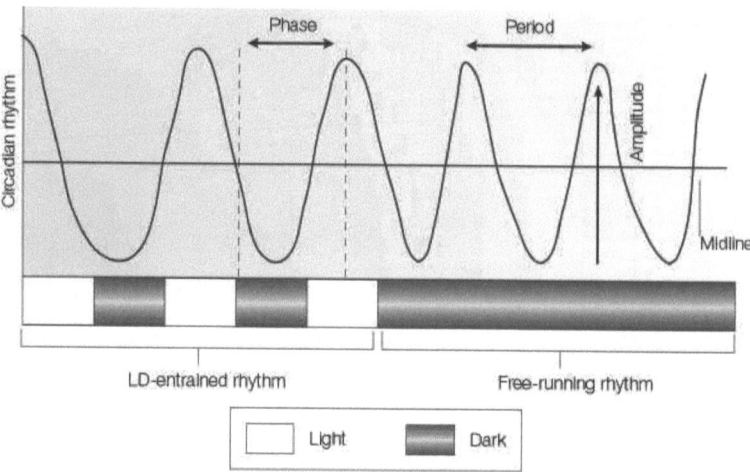

Figure 1.4: Parameters of circadian rhythm. Phase is difference between reference point (e.g., dusk) and the internal rhythm (e.g., the maximum of activity), while the period is length of the internal rhythm. Taken from Bell-Pedersen *et al.*, 2005.

The general properties of circadian rhythms include (Remi *et al.*, 2010):
- Rhythmicity, which is independent of frequency.
- A period in the circadian range, which is determined in constant conditions as free-running period (FRP). In constant dark (or DD conditions) nocturnal species (those active at night) have a FRP less than 24 hours, while diurnal species (those active during day) have a FRP greater than 24 hours (Aschoff, 1960).
- Amplitude robust enough to drive output rhythms.
- Self-sustainment, the rhythm is generated endogenously and can oscillate even in the absence of external stimuli.
- Temperature compensation, which means that period length does not change even if the ambient temperature does (Liu and Bell-Pedersen, 2006).
- Entrainability, which is the adjustment of the clock's period by external signals (*zeitgebers*).
- Circadian rhythms are affected by light intensity. The brightness/intensity or duration of exposure to light can affect the biological clocks. For example, nocturnal species have FRP longer in constant light than in constant darkness, while the opposite goes for diurnal species (Aschoff, 1960).

- Circadian rhythms are under genetic control. For example, mutations in the *period* gene of *Drosophila* may produce phenotypes, which are arrhythmic or express altered periods (Konopka and Benzer, 1971).

Entrainment

Environmental conditions are dynamic and rarely remain constant, so the circadian clocks, to be useful to the organism, have to react and synchronize to the environment. The rhythm has to occur at the correct time of the day because the maintenance of the precise timing with the environment increases fitness (Ouyang *et al.*, 1998). For example, in *Arabidopsis*, it is important that flowering time occurs at the correct phase of the day and season so that the plant can use the maximum sunlight (Michael *et al.*, 2003). This adjustment of the clock is active process known as entrainment (Roenneberg *et al.*, 2003).

Entrainment literally means "to get aboard a train" (originally from the French *entrainer*, "to carry along") (Johnson *et al.*, 2003). This means that an internal biological clock with a free-running period, which is not exactly 24 h, will match its period to the exact 24 h period of the environmental signal. As a consequence entrainment establishes a stable phase relationship between the internal and external oscillations. Entrainment is not same as synchronization, which implies an acute response to the *zeitgeber*, where the period of the internal clock exactly coincides with that of the *zeitgeber*. The complex mechanisms of entrainment mirror the robustness of the clock and its time-of-day-specific responsiveness to *zeitgeber* signals. A clock with a period τ (tau), can be entrained to a *zeitgeber* of period T inducing a phase shift ($\Delta\varphi$): $\Delta\varphi = \tau - T$ (Roenneberg *et al.*, 2003).

Clocks can be entrained by the *zeitgeber* only to a few hours to either side of their natural free-running period (τ). This characteristic of the clock is called range of entrainment, and is species-specific (Roenneberg and Merrow, 2005). Outside of the range of entrainment the clock will not entrain stably, which happens when *zeitgeber* cycle is too short or too long. Entrainment depends on various features of the *zeitgeber* and the internal clock (Remi *et al.*, 2010):

- Structure of the *zeitgeber*: for example the amount of light (photoperiod).
- Strength of the *zeitgeber*: amplitude and intensity.
- Period of the *zeitgeber* cycle (T).
- Period of the endogenous oscillator (τ).
- Dose-response of the circadian system: for example the sensitivity of receptors.

The phase relationship between the clock (φ) and the *zeitgeber* (Φ) is called phase of entrainment ($\psi = \Phi - \varphi$) and depends on the relationship between τ in constant conditions and T. Organisms that differ in this trait are referred to as different chronotypes (Roenneberg *et al.*, 2003).

Traditionally, phase of entrainment has been predicted by the Phase Response Curve (PRC) and Velocity Response Curve (VRC) for a given *zeitgeber* stimulus and τ. However, in the natural environment there are photoperiods (seasonal variation of the day length), which are not easily incorporated into these entrainment models. Therefore, a new model of entrainment was described (Roenneberg *et al.*, 2010). It is based on a Circadian Integrated Response Characteristic (CIRC) that describes how the circadian system integrates light signals at different circadian phases. PRC assumes that light produces advances around dawn and delays around dusk (Hastings and Sweeney 1958; Pittendrigh 1981), while VRC that light accelerates the clock's velocity around dawn and decelerates it around dusk (Daan and Pittendrigh, 1976; Swade, 1969). The CIRC combines these insights so that light around subjective dawn compresses the internal cycle and light around subjective dusk expands it. Shape (determining the extent of their dead zone) and asymmetry (the ratio of its compressing and expanding portions) characterize the CIRCs. By changing the CIRC's shape and asymmetry and an assumed internal cycle length, the entrainment to all *zeitgeber* conditions can be modeled (Roenneberg *et al.*, 2010a).

For example, in humans it has been shown that those who like to go to sleep and get up early tend to have a shorter free-running period than those who prefer to sleep later (Roenneberg *et al.*, 2003). When the cycle length is approximately half of the endogenous free-running period then circadian rhythms often "frequency demultiply" (skip a cycle). Entrainment experiments may also reveal the presence of "masking" effects. Masking is defined as an acute non-circadian effect of a stimulus, which can be induced by a *zeitgeber* signal. Many animals become active or inactive just because light goes on or off. Some of the best protocols to distinguish entrainment and masking are changing *zeitgeber* strength or releasing entrained organisms in constant conditions (Roenneberg *et al.*, 2005).

Relevance of circadian rhythms

As mentioned before, circadian rhythms have roles in different important processes just to name a few: minimizing predations (Fenn and MacDonald, 1995), coordinating mating (Tauber *et al.*, 2003), internal synchronization (Green *et al.*, 2008) and division of labour (Yerushalmi *et al.*, 2006). Therefore, the evolutionary importance of the circadian clock is to anticipate an organism's needs of life, so that they increase the chance of survival and maintain the body in a healthy status (Moser *et al.*, 2006).

The fact that circadian rhythms are found from cyanobacteria to humans suggests their adaptive significance. Several studies using organisms living in constant environments (e.g., a cave-dwelling millipede, blind cave fish) have shown that these organisms possess functional circadian clocks (Koilraj *et al.*, 2000; Espinasa, 2006). This suggests that circadian clocks may have some intrinsic adaptive value and are too beneficial for organisms to be disappear, or that they are simply under a much slower process of extinction than other traits (Jeffery, 2005). Furthermore, studies of wild-type, arrhythmic and long/short period mutants in periodic environments suggest that organisms may have an advantage in those environments, which match their own intrinsic period (Dodd *et al.*, 2005; Woelfle *et al.*, 2004).

The circadian clock influences much of human physiology and behaviour, such as blood pressure, body temperature, hormone secretion (e.g., melatonin, cortisol, testosterone), sleep, etc (Figure 1.5; Foster and Roenneberg, 2008). Therefore, medications can be more or less effective according to the time of day (the study of chronopharmacology). On the other hand, disturbances of the circadian rhythms can affect mood, health and performance (Turek, 2007; Van Cauter and Turek, 1986). Perturbation of clock function has also been implicated in numerous pathologies including circadian sleep disorders, cardiovascular disease, cancer, and metabolic disease (Hastings *et al.*, 2003; Green *et al.*, 2008; Takahashi *et al.*, 2008; Eckel-Mahan and Sassone-Corsi, 2009). Furthermore, many studies have demonstrated that circadian dysfunction is frequent in various neurodegenerative conditions, such as Alzheimer's and Huntington's diseases (for review see Hastings *et al.*, 2008).

There is a particular interest for the effects of internal de-synchronisation. It happens when there is a difference between internal and external time, that means when organisms are not entrained to the external cycle. This de-synchronisation causes chronic health problems (Siegel *et al.*, 1969; Costa, 1996; Rajaratnam and Arendt, 2001; Schernhammer *et al.*, 2001) and may even increase the risk of cancer (Moser *et al.*, 2006). An example of internal de-synchronisation is the Jetlag syndrome that affects time-zone travellers. How long it takes to adjust to a new time zone and recover from jet lag depends on several factors like the number of time zones crossed, the direction of travel, differences among individuals, age, and the particular circadian rhythm involved (Roenneberg and Merrow, 2002). Another example of internal de-synchronisation is shift work. Shift-workers work on evenings or during the night when most of the people rest or sleep. The effect of these disruptions of sleep is often a state of chronic fatigue and sleepiness (Akerstedt, 2003). The worst problems arise for people that change shift times continuously because they cannot get used to the schedule. However, this malaise depends also on the chronotype of the individual: some people adjust better to night schedules because they are "late" chronotypes. The opposite is true for "early" chronotypes, which are more alert in the morning. The individual chronotype is genetically determined but depends on various factors like age, habits, strength of the environmental signals, etc. (Wittmann *et al.*, 2006).

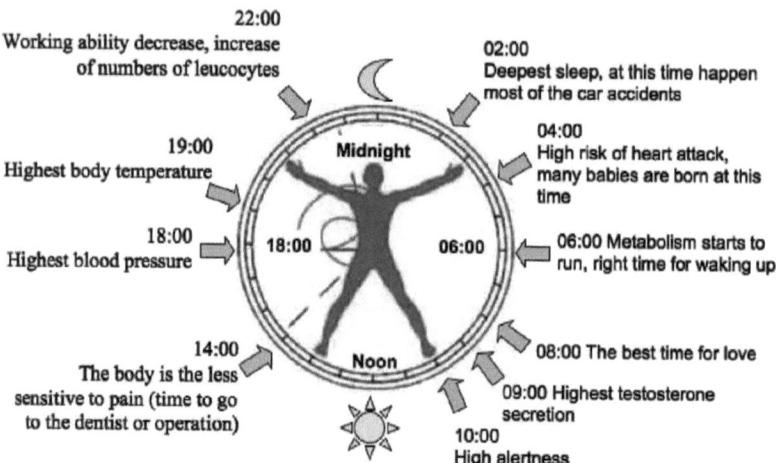

Figure 1.5: The circadian clock of humans. The diagram depicts the various behavioural and physiological variables of someone who rises early in the morning. Figure downloaded on 15.05.2011 from http://circadianrhythmsleepdisorder.info/how-environmental-cues-affect-your-circadian-clock/.

Photoperiodism

The earth's rotation around its axis and its revolution around the sun cause predictable changes in the geophysical environment. The changes in relative duration and magnitude of night and day depend on latitude (Figure 1.6). Hence, these changes can provide temporal information to plants and animals, which use them to anticipate environmental changes and to prepare for coming seasons. This biological phenomenon is called photoperiodism. It is not applicable to regions near the equator where the day length is constant throughout the year, but beyond 5°N or S of the equator, photoperiodism has been well demonstrated (Denlinger, 1986). However, in these regions the seasonal events are also triggered by other environmental inputs, such as temperature change and wet and dry seasons. To be able to respond to photoperiods, an organism must measure the length of day (or night). This requires photoreceptors or other mechanisms capable of detecting the length of light exposure, and to count the number of short days (Hayes et al., 1970; Saunders, 1971).

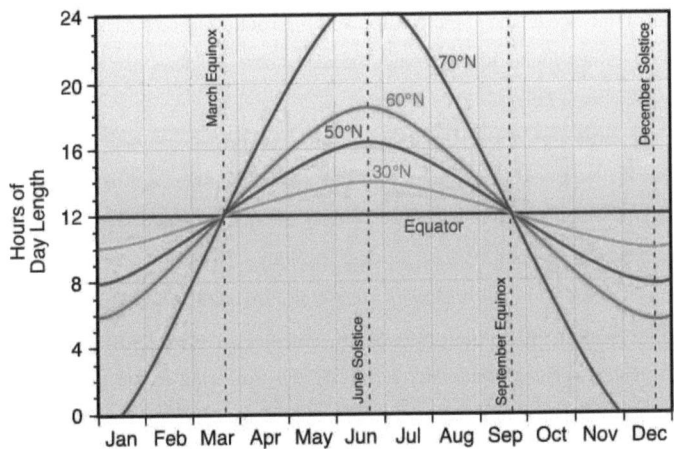

Figure 1.6: Changes of day length throughout the year according to the latitude. Figure downloaded on 15.05.2011 from http://www.physicalgeography.net/fundamentals/6i.html.

Photoperiodism was first discovered in plants in 1920, when Wightman Garner and Henry Allard (Garner and Allard, 1922) performed experiments to test the effect of day length on flowering. They showed that different plants respond to the same day length differently: some species, such as barley, flowered when the day length was longer than a certain critical length, while others, such as soybeans, flowered when the day length is shorter than a certain critical length. The first plants they named long-day plants (LDPs) and the later ones short-day plants (SDPs) (Figure 1.7). Plants, which are not sensitive to the photoperiod, are called day-neutral plants.

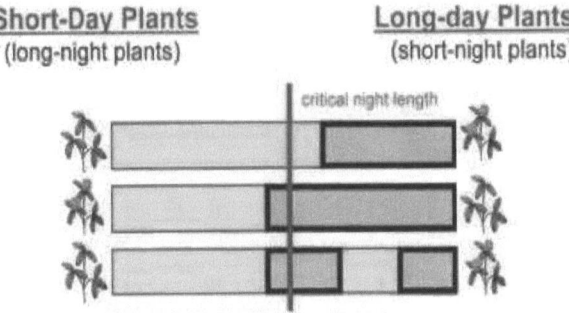

Figure 1.7: Schematic representation of different behaviour in short-day and long-day plants. Short-day plants flower when the length of night is above the critical night length, while long-day plants flower when the night length is below the critical night length. Figure downloaded on 15.05.2011 and redrawn from http://www.ext.colostate.edu/mg/gardennotes/142.html.

The geographical distribution of many plants is influenced by photoperiod. For example, ragweed (a SDP) is not found in northern Maine because it needs the day length shorter than 14.5 h to flower. Since in northern Maine, days do not shorten to this length until August, this is too late for the maturation of the seeds and therefore the species cannot survive there. On the other hand, spinach (a LDP) is not found in the tropics because there the days are never long enough for flowering (Raven et al., 1999).

The length of the critical photoperiod varies not only between species but also between the same species at different latitudes. For example, in the pitcher plant mosquito *Wyeomyia smithii* the critical day length lengthens systematically toward the north (Bradshaw, 1976). Most critical photoperiods are between 10 and 14 hours of light (Binkley, 1997).

Photoperiodism is found in many organism from fungi, invertebrates to humans and is seen as a behavioural, molecular and physiological response essential for coping with changing day lengths. Photoperiod provokes different responses, such as: the flowering in the plants (Franklin et al., 2005), migration and diapause in insects (Reppert, 2006; Kostal, 2006), changes in the colour of fur and feathers (Kauffman et al., 2001; Jovani et al., 2010), bird's migration (Gwinner, 1977), entry into hibernation, sexual behaviour (Pohl, 1987), and the resizing of sexual organs (Rowan, 1925). In humans, reproduction is also subjected to seasonal variation. Human conceptions exhibit two peaks per year, one in the spring and one in the fall (Roenneberg and Aschoff, 1990). In addition, reproducible hormones exhibit significant seasonal variation in men (Levine et al., 1994) and women (Kivela et al., 1988).

Models of daylength measurements

Although the photoperiodism has been studied at the behavioural and physiological level, the molecular mechanisms are still unknown. There are three models trying to explain the molecular mechanism of photoperiodism: the hourglass model, external coincidence model and internal coincidence model.

In the hourglass model, which was proposed by Anthony D. Lees (1973), the organisms monitor the accumulation of some "physiological agent" during one part of the light-dark cycle (Figure 1.8). This process is reversed during the other portion of the cycle. The absolute duration of light or dark is monitored and if the light (or the dark) is long enough, a threshold is reached and a response is initiated. The hourglass model is used by several invertebrates (Veerman and Vaz, 1987; Saunders and Bertossa, 2011) but only by very few vertebrates (Underwood and Hall, 1982). In this model circadian clock plays no role because the hourglass has no endogenous rhythmicity and must be reset each day.

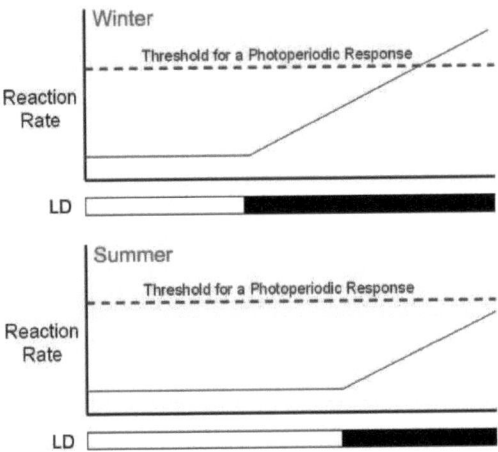

Figure 1.8: Hourglass model of photoperiodism. The response is initiated when the threshold of "chemical" is reached (red dashed line), for example when the night is long enough. In summer the nights are shorter and therefore the "chemical" is degraded. Downloaded on 15. 05. 2011 from http://www.colorado.edu/intphys/Class/ IPHY3730/15 photoperiodism.html.

Erwin Bünning proposed the external coincidence model in 1936 (Figure 1.9). This model proposes that night-phase is sensitive to light while the day-phase is photo-insensitive. The circadian clock generates a rhythm of photoperiodic photosensitivity (CRPP) and when light starts illuminating the photosensitive phase it triggers the physiological or behavioural response. In this model light has a dual effect: it entrains the rhythm of photosensitivity and is directly required as a stimulus. The name "external coincidence" model derives form the fact that external stimulus (light) has to coincide with that of an internal rhythm of sensitivity to light.

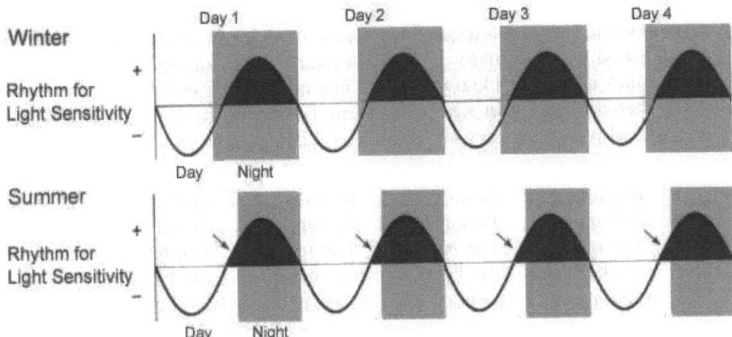

Figure 1.9: External coincidence model of photoperiodism. Light, an external cue, actively induces the photoperiodic response (arrows) when the photosensitive phase of the cycles is illuminated. Figure downloaded on 15. 05. 2011 from http://www.colorado.edu/intphys/Class/IPHY3730/14biologicalrhythms.html.

In contrast, in the internal coincidence model (Figure 1.10), proposed by Colin Pittendrigh and

Dorothea Minis in 1964, the light's role is to entrain the circadian system. This model assumes the existence of two or more internal oscillators that are normally out of phase, but are brought to internal coincidence by the re-phasing of one or both by light occurring at appropriate time. Studies of photoperiodism in insects supported this model (Vaz Nunes and Saunders, 1999).

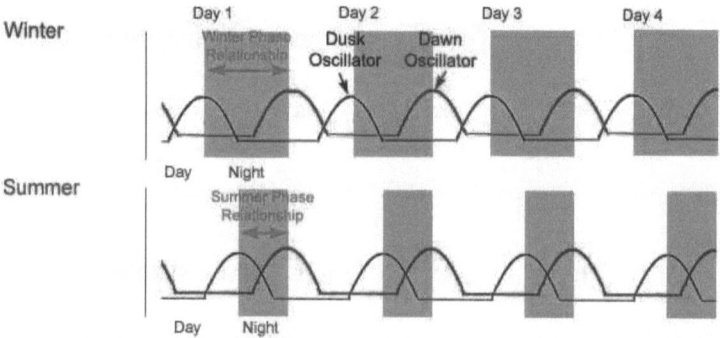

Figure 1.10: Internal coincidence model of photoperiodism. Light only sets the phasing of the dusk and dawn oscillators. Figure Downloaded on 15. 05. 2011 from http://www.colorado.edu/intphys/Class/IPHY3730/ 14biological rhythms.html.

Neurospora crassa

Fungi comprise probably the most biotechnologically useful group of organisms and are used to synthesize a wide range of important compounds, like enzymes, antibiotics and secondary metabolites, and therefore they have taken part in the progress of biochemistry, genetics, and molecular biology. The kingdom *Fungi* is divided into six phyla (Bruns, 2006): *Chytridiomycota*, the most primitive fungi, *Zigomycota*, *Basidiomycota*, *Ascomycota*, *Glomeromycota* and *Microsporidia* (Figure 1.11). All fungi are eukaryotic organisms having two common features: they grow vegetative through a mycelium and their nutrition is based on absorption of the organic matter.

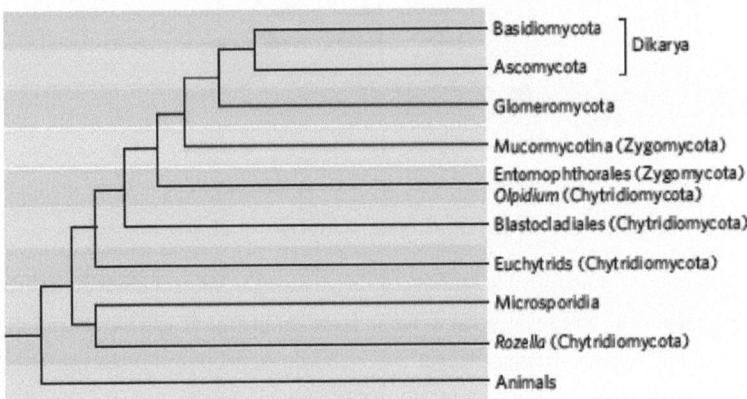

Figure 1.11: The branches of kingdom *Fungi*. Taken from Bruns, 2006.

Neurospora crassa is a filamentous fungus belonging to *Ascomycota* in which the formation and maturation of specialized cells, the ascospores occur. A typical filamentous fungus is made of a mass of branched, tubular filaments enclosed by a rigid cell wall. The filaments (called hyphae) branch into a complicated network called the mycelium. The mycelium makes up the body of the fungus, the thallus. Fungal growth is mainly confined to the tips of the hyphae (Davis, 2000).

Neurospora was known as "red bread mould" and was widely studied in 1850 for its diffusion as contaminant of French bakeries. There are 35 species of *Neurospora* and they are divided in three groups based on sexual characteristics (Nygren *et al.*, 2011):

- Heterothallic: *Neurospora crassa, discreta, intermedia, sitophila, metzenbergii, hispaniola, perkinsii.*
- Pseudohomothallic: *Neurospora tetrasperma, tetraspora.*
- Homothallic: *Neurospora africana, dodgei, galapagosensis, lineolata, terricola, pannonica, brevispora, haspidophora, udagawae, indica, reticulospora, stellata, cerealis, minuta, kobi, pseudoreticulata, uniporata, retispora, santi-fiori, novoguineensis, dictyophora, saitoi, calospora, endodonta, sublineolata, nigieriensis.*

Neurospora crassa was first described by Shear and Dodge (1927) who characterized the sexual biology of two eight-spored heterothallic species and one four-spored apparently homothallic species. They named the genus *Neurospora* because of the nerve-like ornamentation on developing ascospore walls (Figure 1.12).

Figure 1.12: The ascospores of *N. crassa* from which the genus name derived (photo by N. B. Raju).

Neurospora crassa was first used to understand the relationship between proteins and genes. It is one of the organisms studied which led to the formulation of Beadle and Tatum's so called "one gene-one enzyme" hypothesis. Since then it was used as a model organism for numerous genetic, cytogenetic, biochemical, molecular and population biology studies, and also for studying the regulation of carotenoid biosynthesis and photobiology.

Neurospora crassa has many advantages, which makes it a well-suited model organism in biology. First of all, it is non-pathogenic and spends most of its life cycle as a haploid organism. This means that gene expression will not be influenced by dominance or recessive alleles, and mutations will not be masked. Next, *Neurospora* has a fast reproduction requiring only about two weeks and the ascospores are large enough for manual isolation. Nutritional requirements are very simple, it can easily be grown in large quantities and is inexpensive to maintain. In addition, the vegetative sporulation exhibits a clearly defined clock output when growing on solid media. Manipulation of the cultures is very easy and requires only standard microbiological techniques. Finally, the growth of its cultures across an agar surface can reflect differences in strains and environmental conditions, such as temperature and nutritional richness of the medium (Feldman and Hoyle, 1973).

The *Neurospora* genome has been completely sequenced. It consists of only 43 megabases and has approximately 10 000 genes on seven chromosomes (Galagan *et al.*, 2003). It has a large number of genes without homologues in the yeast *Saccharomyces cerevisiae*. This, together with its multicellularity, makes *Neurospora* a good model system for higher eukaryotes (Borkovich *et al.*, 2004). Many mutants have been characterized and can be used to distinguish cell types during the vegetative and sexual phases of the life cycle. All mutants, knock-out and wild-collected strains are available and can be easily ordered online (Fungal Genetic Stock Center: http://www.fgsc.net/). Furthermore, it displays a number of gene-silencing mechanisms acting in the sexual or the vegetative phase of the life cycle (Borkovich *et al.*, 2004). It is an obligate aerobe and being a heterotrophic organism it has oxidative phosphorylation. Cultures of *Neurospora* remain viable for decades, stocks can be frozen at -20°C from year to year, and then quickly restored. There are two distinct mating types for making genetic crosses. During the sexual cycle the spores are kept in order reflecting the arrangement of homologous chromosome pairs. Transformation is efficient where 100% homologous recombination can be achieved, allowing targeted gene disruption using a variety of selectable markers (Aronson *et al.*, 1994a).

Life cycle

Neurospora crassa's life cycle is haploid: the only diploid stage is the zygote. Individuals are always haploid, even so the hyphae are coenocytic, i.e., they are multinucleate cells where the nuclei are not separated by cell walls. Mycelia can be heterokaryons, i.e., cells containing multiple, genetically different nuclei. *Neurospora* has two kinds of reproduction: asexual and sexual (Figure 1.13), both are regulated by the clock (Davis, 2000).

The asexual cycle starts with germination of aerial hyphae leading to haploid asexual spores (conidia). The multinucleate macroconidia and uninucleate microconidia are then dispersed into the environment and, if they land on a suitable substrate, the asexual cycle starts again (Davis, 2000).

The sexual cycle involves haploid nuclei of two different mating types. The mating-types are defined by alternative forms of the genetically complex mating type-region: *mat* A and *mat* a (Perkins and Turner, 1988). In conditions of reduced nitrogen and carbon, the "female" individual forms a multi-cellular protoperithecium, a sexual structure, which has a specialized hypha called trichogyne. This trichogyne can fuse with a conidia or a hypha of the opposite mating type. When the fusion takes place, the nucleus of the conidia travels through trichogyne and hits the ascogonium, a cell of protoperithecium that acts as the female gamete. Here the two different nuclei divide many times forming an ascogenous hypha, and then nuclear fusion takes place. Each diploid nucleus divides through two meioses: the four meiotic products undergo one further mitotic division. The resulting eight nuclei are closed in a rigid wall, which constitute the haploid ascospores. After seven days, the ascus breaks open, and the spores are forcibly ejected. Ascospores can survive for long periods and are activated by heat. Germination of ascospores represents the initiation of the next asexual cycle (Davis, 2000).

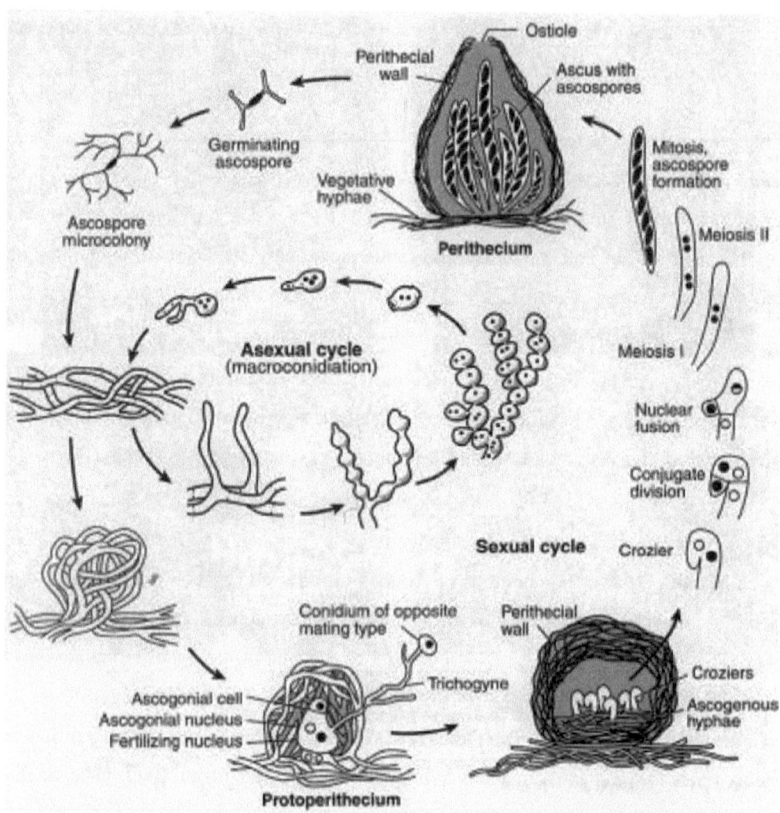

Figure 1.13: Life cycle of *Neurospora crassa*. The asexual cycle consists of conidiation, the production of asexual haploid spores (conidia) that will form a new mycelium. In the sexual cycle, the fusion of nuclei from different mating types produces the haploid ascospores that also will form a new mycelium. Taken from Davis, 2000.

Natural habitat

The kingdom *Fungi*, or *Mycota* - including yeasts, rusts, smuts, mildews, moulds, and mushrooms - comprises about 1.5 million species and are among the most widely distributed organisms on earth (Müller and Schmit, 2007). They can be found free-living in soil or water (fresh or marine water), or form parasitic or symbiotic relationships with plants or animals, respectively. Most of them live in temperate and tropical regions of the world, where is sufficient moisture to enable their growth. However, they have been reported also in the Arctic and Antarctic regions, where they constitute lichens, in the symbiosis with algae. In general, fungi are abundant in moist habitats where organic matter is plentiful and are less abundant in drier areas or in habitats with little or no organic matter. Sexual reproduction, an important source of genetic variability, allows the fungus to adapt to new environments (Pandit and Maheshwari, 1994).

More than 5 000 isolates of *Neurospora* species have been catalogued from worldwide collections (Turner *et al.*, 2001). *Neurospora crassa* isolates have been found at different southern and northern latitudes, most of them collected in moist tropical or subtropical climates, but they have also been found in temperate forests in North America and Europe (Jacobson *et al.*, 2004; Jacobson *et al.*, 2006). Species diversity is based on geographical distribution and latitude (Figure 1.14; Jacobson *et al.*, 2006).

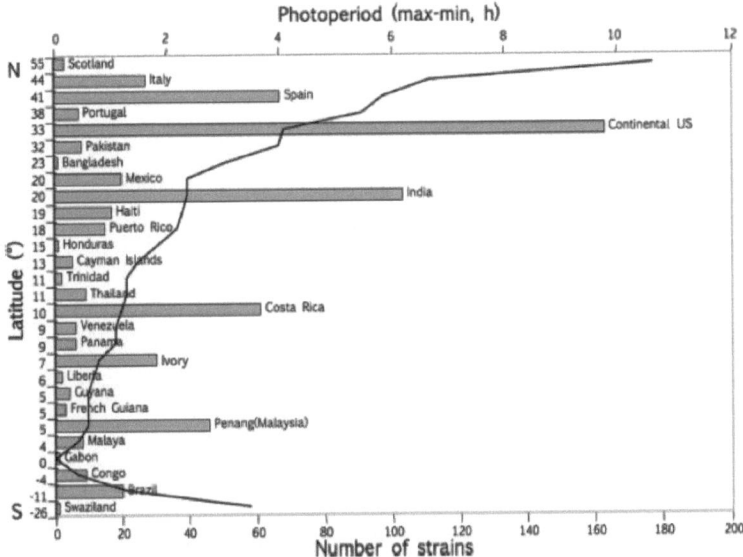

Figure 1.14: Latitudinal and geographical distribution of the wild-type isolates of *Neurospora crassa* deposited in the Fungal Genetic Stock Center, Kansas (source: http://www.fgsc.net; Turner *et al.*, 2001; Jacobson *et al.*, 2006). The respective photoperiodic differences between longest and shortest day of the year are drawn as a line (upper horizontal axis).

In its natural environment, *N. crassa* can be found on burned vegetation killed by fire, like woody and herbaceous plants and beneath the bark of the trees. *Neurospora* is the first colonizer after fires have destroyed most of the vegetation, living on the remaining carbon sources. The importance of fire is twofold: firstly, fire produces a sterile environment rich in nutrients, and secondly, fire provides heat necessary for ascospore germination (Jacobson *et al.*, 2004). *Neurospora* shows annual rhythms in the change of the concentration of spores, which are correlated to environmental conditions, such as available nutrients, humidity, wind speed, or temperature (Ingold, 1971).

Evolution

Until the 1960s, fungi were considered as members of the plant kingdom, but when the five-kingdom system of biological classification was introduced, they were classified into a single one

(Whittaker, 1969). Recent studies based on comparison of the amino acid sequences from fungi, plants, and animals indicate that fungi are evolutionary closer to animals than to plants (Baldauf and Palmer, 1993).

So far the phylogeny of the fungi was based on sequencing of the single gene encoding for a small subunit of ribosomal RNA, but this is not sufficient to create a "strong" evolutionary tree. A single-gene tree is always questionable, because different genes can give different views of evolutionary history (Bruns, 2006). Therefore, the evolutionary relationships of the fungi are not well understood. However, broad sampling of wild-collected isolates and fully sequenced genome makes *Neurospora* an ideal model organism to investigate evolution, speciation mechanisms, and genetics of reproductive isolation in fungi (Turner *et al.*, 2010).

The determination of a new *Neurospora* species has been based on morphological, biological and phylogenetic species recognition (Taylor *et al.*, 2006). The morphological species recognition (MSR) is based on morphological or other phenotypic characters, e.g., growth, production of secondary metabolites or the presence of pigments; while the biological species recognition (BSR) uses mating tests to reveal the fertility of the crosses, and therefore the reproductive isolation. Although the species of a strain cannot be determinate reliably based on MSR, the BSR method is traditional used to classify different *Neurospora* species (Perkins *et al.*, 1976). It works very well in most cases but has some disadvantages, as it does not work for asexual species or in the species in which sexual reproduction is not easily induced in laboratory. Thus, BSR fails to distinguish among reproductively isolated species (Dettman *et al.,* 2003a). However, the phylogenetic species recognition (PSR) uses genealogical concordance of DNA sequence of appropriately polymorphic loci. Thus, PSR can be applied even to asexual or uncultivable organisms and is less dependent on prior knowledge of the existence of a species (Dettman *et al.*, 2003). In *Neurospora crassa*, this method discovered three different subgroups, named clade A, B, and C. Clade A was found predominantly in the Caribbean Basin and the Ivory Coast, clade B in Europe, India and the Ivory Coast, and clade C only in India (Jacobson *et al.*, 2006).

The difficulties of *Neurospora* phylogeny are principally due to the wide distribution of this genus, where in most cases natural barriers prevent gene flow (pre-zygotic reproduction isolation). Reproductive isolation caused by natural barriers can bring to allopatric speciation, which is the predominant speciation among the species of *Neurospora* (Dettman *et al.*, 2003). Moreover, to adapt to different environmental conditions genetic differences organisms can evolve through post-zygotic isolation. Post-zygotic isolation is manifested by an imperfect development or a sterility of the hybrid. The natural selection against maladaptive reproduction can increase the evolution of isolation: this phenomenon is called reinforcement (Ridley, 2003). Evidence of reinforcement in *Neurospora* can be seen as decrease of production of ascospores and less viable ascospores from inter-specific mating (Dettman *et al.*, 2003a).

Circadian systems, because of their universality and ancient origin, are ideal for evolutionary studies to assess the relationship between organisms. For example, the *frequency* gene (see below) can be useful to investigate evolutionary relationship between closely and distantly related species of ascomycete fungi due to an irregular pattern of conserved regions interspersed with variable regions in the gene (Lewis and Feldmann, 1996). In other example, simple sequence repeats (SSRs) can be used to detect variation between organisms. Although the functional role of tandem repeats is poorly understood, they can be successfully used as genetic markers (Borstnik and Pumpernik, 2002). SSRs are made by mechanisms involving slipped-strand mis-pairing and unequal crossing-over (Levinson and Gutman, 1987). There are three SSRs found in *Neurospora crassa* of White collar-1 protein (WC-1, see below): 5'AG/GA, in the amino-terminal polyglutamine (NpolyQ) domain, and carboxly-terminal polyglutamine-histidine (CpolyQH) domain. It has been shown that variation in NpolyQ domain is correlated with the circadian phenotype and the environment of the isolates (Michael *et al.*, 2007).

Circadian clock

Fungi provide a powerful model to study the molecular basis of the circadian rhythms, because their circadian system is simple and without complex multi-cellular interactions (Bell-Pedersen *et al.*, 1996). The circadian system of *Neurospora crassa* has been one of the best described at the physiological and molecular level.

At the physiological level, the asexual spore development (conidiation) is used as an output of the circadian clock and has proven to be useful in measuring the effects of mutations on clock (Liu and Bell-Pedersen, 2006). The circadian rhythm of conidiation is easily assayed in race tubes. In the *band* (*bd*) strains it can be seen as a series of conidiation (spore-forming) regions or "bands" on the surface of an agar medium. There any point of the conidiation rhythm can be used as phase reference but onset has proven to be the most reliable marker for phase of entrainment (Roenneberg *et al.*, 2005).

Neurospora crassa can be entrained by different light-dark and temperature cycles and temperature pulses can reset the rhythm. Its period is temperature compensated between 18°C and 30°C (Loros and Feldman, 1986). Both light and temperature have similar entrainment effects on the circadian clock in that 12/12 h cycles of light-dark or warm-cold result in a 24 h rhythm of conidiation with the peak of conidiation occurring just prior to the transition from dark to light or from cold to warm (Merrow *et al.*, 1999). When the standard laboratory strain *bdA* is grown in constant darkness conidiation occurs every 22 h (Sargent *et al.*, 1966). The circadian clock of *Neurospora* is extremely light sensitive as moonlight levels can affect conidial banding (Sargent *et al.*, 1966), while in constant light, the banding pattern stops (Pittendrigh *et al.*, 1959).

All known light-induced responses in *Neurospora* are regulated by the blue light such as: (1) the phase shift in the circadian rhythm of conidiation (Sargent and Briggs, 1967), but also (2) induction of carotenoid synthesis in mycelia (Harding and Turner, 1981), (3) induction of conidiation (Lauter *et al.*, 1997), (4) protoperithecium formation under nitrogen-limited conditions (Degli Innocenti and Russo, 1984), (5) light-dependent perithecial polarity (Oda and Hasunuma, 1997), and (6) positive phototropism of perithecial beaks (Harding and Melles, 1984).

The circadian oscillator of *N. crassa* is composed of self-sustaining cellular feedback loops where almost all responses require the protein products of three genes, the *frequency (frq)*, *white collar-1 (wc-1)* and *wc-2* (Lee *et al.*, 2000). The *Neurospora* clock starts to cycle from late subjective night. At this point, most of the FRQ protein in the cell has been degraded, and *frq* mRNA levels are low (Dunap and Loros, 2004). WC-1 and WC-2 are transcription factors, which have the two PAS domains (PER, ARNT, and SIM, a zinc-finger DNA-binding domain) and glutamine-rich putative transcription activation domain (Cheng *et al.*, 2002). They form a heterodimeric complex, the White-collar complex (WCC), through their PAS domains (Liu, 2003). WCC is localized in the nucleus, and binds to the C-box of the *frq* promoter, activating its transcription (Crosthwaite *et al.*, 1997). By early morning and after splicing of the *frq* mRNA, FRQ protein is produced. After the FRQ protein is synthesized, it dimerizes with itself (through the N-terminal coil-coiled domain) and forms a complex with FRH (FRQ – interaction RNA helicase) called FFC (Cheng *et al.*, 2005). FFC in the nucleus represses the transcription of *frq* by inhibiting WCC activity (Froehlich *et al.*, 2003). The consequence is a decrease in *frq* levels with the maximum around the subjective midnight. In the cytosol FRQ protein is progressively phosphorylated by several kinases and dephosphorylated by two phosphatases (Liu *et al.*, 2000). When FRQ becomes extensively phosphorylated, it interacts with FWD-1 (an F-box/WD-40 repeat-containing protein) and this brings to the ubiquitination and degradation of FRQ by the proteasome system (He and Liu, 2005). When the levels of FRQ decrease under a certain threshold, WCC is no longer inhibited by FFC, and *frq* transcription is reactivated around the subjective late night to start a new cycle. The results are a rhythmic *frq* RNA and FRQ protein levels (Figure 1.15).

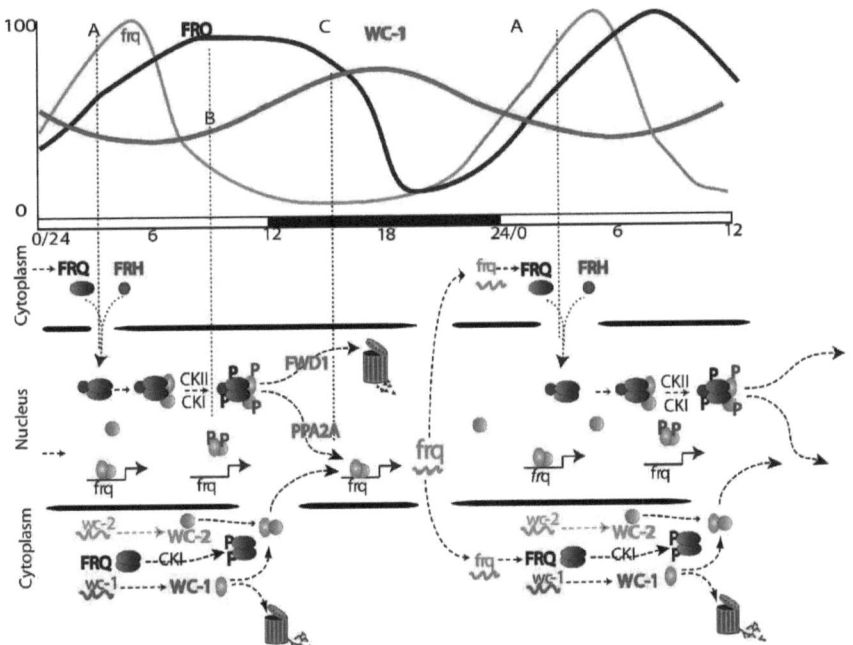

Figure 1.15: Transcription/translation feedback loop in *Neurospora*. The top shows how the levels of *frq* mRNA, FRQ, and WC-1 proteins oscillate through time over one and a half cycles in constant darkness (subjective day light bar and subjective night dark bar are shown). Below this, different proteins move between the cytoplasm and nucleus in the circadian cycle. At time A in the morning, WC-1 and WC-2 are driving *frq* expression, and FRQ is actively translated, associates with FRH, moves to the nucleus, and associates with and leads to (around time B, afternoon) the phosphorylation of WC-1 and WC-2. This phosphorylation inactivates WC-1 and WC-2 so *frq* expression drops. Around the same time FRQ in the cytoplasm helps newly made WC-1 and WC-2 to associate. Moving toward time C, phosphorylated FRQ associates with FWD-1 and goes to proteasome (trash can) allowing old WC-1 and WC-2 to be dephosphorylated by PP2A and reactivated; new WC-1 and WC-2 move to the nucleus and join in reactivation of *frq* transcription in the late night. Taken from Dunlap, 2006.

This *frq/wc (frequency/white-collar)*-based circadian oscillator (FWO) displays similarity with the feedback loops found in higher eukaryotes, supporting a *Neurospora* as a good model organism to investigate the clock (Liu and Bell-Pedersen, 2006). However, the circadian system may include slave oscillators beside the FRQ/WCC circadian feedback loop, such as simple clock-regulated genes and proteins (Nowrousian *et al.*, 2003). These slave oscillators are named FRQ-less oscillator (FLO) (Iwasaki and Dunlap, 2000). The FLO does not show a circadian entrainment but it is directly influenced by environmental factors. Rhythms controlled by FLO are not robust, as phase and period are variable and imprecise, and as they have not temperature compensation (Pregueiro *et al.*, 2005).

WC-1 and WC-2 proteins have two different roles in *Neurospora* (Crosthwaite *et al.*, 1997). First, they are both important for the light induction of gene expression of all known light-induced genes, and second, in constant darkness, WC-1 and WC-2 are necessary for generation of the circadian rhythms by activating the *frq* transcription (Cheng et al., 2001). The WC-1 protein was also identified as a blue-light photoreceptor (Froehlich et al., 2002). It contains a chromophore-binding motif called the light, oxygen, and voltage (LOV) domain. If this domain is removed from WC-1, many of the light-regulated responses of WC-1 are lost (Cheng et al., 2003). In constant darkness, neither *wc-1* nor *wc-2* RNA oscillate, but WC-1 protein content shows a clear rhythm (Lee et al., 2000), although this is not essential for the working of the clock. The rhythms of WC-1 and of FRQ are out of phase (Lee et al., 2000) and physical interaction between FRQ and WCC are necessary for closing the loop (Hong et al., 2008). Several other loops, such as positive action of FRQ on *wc-2* expression (Cheng et al., 2001), the positive regulation of WC-1 by WC-2, and the repression of *wc-2* expression by WC-1 (Cheng et al., 2002), drive the clock. WC-2 is predominantly in the nucleus and does not show a circadian regulation but it is required for the interaction between WC-1 and FRQ (Denault et al., 2001). Data demonstrated that WC-1 is the limiting factor in the WCC complex, suggesting that with greater WC-1 and WC-2 levels, the level of the FRQ oscillation is higher and therefore the overall rhythm is more robust (Cheng et al., 2001).

Furthermore, it was seen that the *frequency* gene is essential for temperature entrainment: temperature cycles with amplitudes of 2°C are sufficient to entrain the rhythm (Nowrousian et al., 2003). Two distinct FRQ proteins can be translated from *frq* mRNAs, a large form of 989 aminoacids (LFRQ) and a smaller form of 890 aminoacids (SFRQ), both can interact with the WCC (Garceau et al., 1997). Temperature influences the forms of FRQ protein synthesized, their levels and activities: a threshold level of FRQ increases with increasing temperature (Aronson et al., 1994). The ratio of long to short FRQ protein is regulated by thermo-sensitive splicing of intron 6 of *frq* gene (Diernfellner et al., 2005) and is responsible for temperature compensation of the clock (Aronson et al., 1994). Both forms are needed for optimally robust rhythmicity. For example, short and less FRQ is required at lower temperatures (<~22 °C), while the large form and higher levels of FRQ are needed at higher temperatures (>26 °C) (Liu et al., 1997). However, there is little knowledge of the molecular mechanisms of *Neurospora*'s entrainment to temperature (Ruoff et al., 2005).

Many mutations in clock genes are known that affect the circadian rhythmicity, and molecular analyses of these genes have contributed to model the circadian oscillators. Mutations in the FRQ protein, for example, affect temperature compensation of the circadian rhythm and sensitivity to light-induced phase resetting (Aronson et al., 1994). Both short- and long-period alleles are known at the *frq* locus (Lakin-Thomas et al., 1991). To name just few: *frq-1* is a short-period mutant (FRP = 16.5 h), while *frq-7* is a long-period mutant (FRP = 29 h), which has lost part of its temperature-compensating ability. Both mutants are the result of single point mutation (G to A transition) in the *frequency* gene (Merrow and Dunlap, 1994). There are two null mutations at the *frq* locus: *frq-9*, which has a single base pair deletion producing a truncated protein, and *frq-10*, which is a deletion mutation constructed by targeted gene disruption (Loros *et al.*, 1986; Aronson *et al.*, 1994). These two mutations show identical phenotypes: in the first few days of growth on race tubes they show arrhythmic but later produce rhythmic banding. This suggests that the entrainment in *Neurospora* is not only given by the known transcription/translation feedback loop. The *white-collar* mutants, *wc-1* and *wc-2*, are called "blind" mutants because they are insensitive to the blue light (Perkins *et al.*, 1982). *White-collar* mutants are conditional rhythmic, meaning they produce bands of conidiation under some culture conditions and rhythmicity cannot be induced by light or temperature steps (Dragovic *et al.*, 2002).

Aims of the study

Neurospora is an organism found in different geographical locations around the world (Jacobson *et al.*, 2004) and therefore is a good model system for studying photoperiodic response. The aim of this thesis was a phenotypic and genotypic comparison between 24 wild-type isolates of the species *Neurospora crassa*. These isolates come from different latitudes and thus environmental conditions (Figure 1.16) and were used the answer the question: is genetic variation in the clock genes and circadian phenotype selected by environment in *Neurospora crassa*?

To try answer this question, the work was divided in two parts. In the first part, four neutral markers, two regions of the *frequency* promoter and *white collar-1* gene were sequenced. With help of neutral markers, the phylogenetic species were found and used to test if differences found at *frq* promoter and *wc-1* are due to phylogenetic groups. *Frq* promoter and *wc-1* were sequenced to search for single nucleotide polymorphisms (SNPs), which were used to see if the groups found in phylogenetic analysis correlate with latitude of the origin. Secondly, all isolates were tested in different light-dark and temperature cycles performing "circadian surface" experiments. These experiments permitted to see how different isolates behave in different conditions and if their responses correlate with latitude of the origin. Finally, from this two parts popped out the question: does circadian physiology reflect genetic differences?

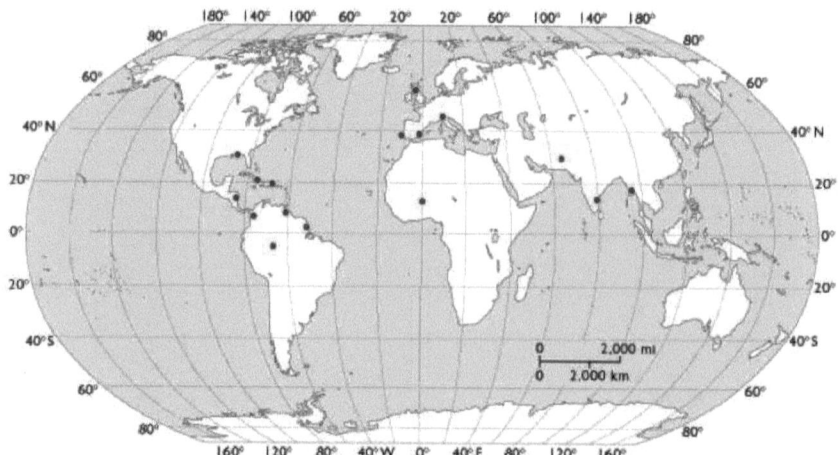

Figure 1.16: Collection sites of *N. crassa* isolates used in this thesis. Map downloaded on 29.08.2011 from: http://www.roebuckclasses.com/maps/placemap/placemapindex.htm

2. MATERIALS AND METHODS

Strains

Twenty-six wild-type isolates of *Neurospora crassa* from 16 countries of different latitudes and longitudes were investigated (Table 2.1), which have been generously provided by Prof. David J. Jacobson (University of California, Berkeley). Thailand and Pakistan isolates were used only for sequencing of neutral markers.

In addition, the standard laboratory strain *bdA* (FGSC #1858) was used as control in all physiological experiments. It has a free-running period (FRP) of 21.5 h and carries the *band* (*bd*) mutation, which allows clear expression of the circadian conidiation rhythm (Sargent *et al.*, 1966). However, it has no effect on phase or period length and prevents inhibition of conidiation by high CO_2 concentrations in closed culture vessels. The strain was obtained from Fungal Genetic Stock Center (http://www.fgsc.net/). Furthermore, *Neurospora tetrasperma* (FGSC #2508) was used as out-group strain in the sequencing experiments. The sequence was obtained from DOE Joint Genome Institute (http://genome.jgi-psf.org/Neute1/Neute1.home.html).

Table 2.1: The list of strains used in this thesis. FGSC = Fungal Genetic Stock Center

FGSC #	Collection site	Latitude	Longitude	Mating type	Clade*
10867	Scotland UK, Edinburgh	55° 57' N	3° 13' W	A	B
10036	Italy, Turchino Est	44° 27' N	8° 44' E	a	B
10866	Italy, Genova	44° 26' N	8° 45' E	a	A
10056	Italy, Genova	44° 26' N	8° 45' E	A	B
10860	Spain, Seros	41° 23' N	0° 10' E	A	B
10861	Spain, Seros	41° 23' N	0° 10' E	a	B
10862	Spain, Seros	41° 23' N	0° 10' E	a	B
10043	Spain, Seros	41° 23' N	0° 10' E	a	B
10863	Spain, Seros	41° 23' N	0° 10' E	A	B
10864	Spain, Seros	41° 23' N	0° 10' E	A	B
10045	Spain, Seros	41° 23' N	0° 10' E	a	B
10865	Spain, Seros	41° 23' N	0° 10' E	a	B
10046	Spain, Seros	41° 23' N	0° 10' E	A	B
10024	Portugal, Tapada de Mafra	38° 58' N	9° 17' W	A	B
1825	Pakistan, Lahore	31° 34' N	74° 22' W	a	?
8877	Louisiana USA, Franklin	29° 48' N	91° 31' W	A	A
3693	Puerto Rico, Colonia Paraiso	18° 29' N	66° 8' W	A	A
8816	Haiti, Carrefour Dufort	18° 27' N	72° 38' W	A	A
6797	Thailand, Khao Eto	14° 05' N	101° 23' W	A	?
8859	India, Mallilinatham, Tamil Nadu	12° 40' N	80° 05' E	A	C
6211	Costa Rica, Jaco	10° 0' N	85° 50' W	A	A
8828	Ivory Coast, Tissale	5° 53' N	4° 57' W	A	A
5914	Guyana, Torani Canal	5° 45' N	57° 30' W	A	A
6233	Venezuela, Puerto Ayachucho	5° 39' N	67° 32' W	a	A
7553	French Guiana, Devils Island-Ile St. Joseph	4° 0' N	52° 30' W	a	A
4705	Brazil, Rondon	1° 47' N	63° 0' W	A	A
1858	bdA, standard laboratory strain			A	

* normal font from Dettman et at., 2003; italic from Jacobson et al., 2006; bold from this study. In same color are isolates from same country.

Strain maintenance

All isolates were kept in slants with Vogel's Minimal Medium (Vogel, 1956; Appendix) at room temperature. The slants were plugged with cotton in order to avoid contaminations. They were inoculated from the original stocks at the day of arrival, and then allowed to grow on the bench for seven days before using or being wrapped individually with Parafilm for long-term storage. For storage, all strains were preserved at –20°C. To avoid the accumulation of background mutations, no further subcultures of these freezer stocks were made. 4 - 5 days after inoculation, conidia could be used to inoculate other slants, flasks, race tubes and liquid cultures. Although conidia production stops after 7 - 10 days, they usually stay viable at room temperature for a year or longer (Perkins, 1973).

Molecular methods

DNA preparation

Isolates were grown in 100 ml Erlenmeyer flask with Minimal Medium (Appendix) with 2% glucose for 2 - 3 days at room temperature in an orbital shaker under constant light. Mycelial tissue was dried between paper tissues, submerged in liquid nitrogen and ground into a powder. Dry tissue was incubated at 65°C for 1 h in 600 µl of lysis buffer with final concentrations of 100 mM Tris-HCl, 50 mM EDTA, 1% SDS, and 20 mg/ml Proteinase K (BioLabs, Frankfurt, Germany). 7.5 M ammonium acetate was added and samples were centrifuged at 13800 rpm on 4°C for 3 min. The supernatant was incubated with RNase A (10 mg/ml; Roche Diagnostics, Mannheim, Germany) for 1 h at 37°C. After a wash with chloroform-isoamyl alcohol (24:1), samples were centrifuged (13800 rpm for 8 min) to remove cellular debris. The aqueous phase was collected and genomic DNA was extracted using isopropanol by centrifuging at 13800 rpm for 30 min at 4°C. The pellet containing genomic DNA was washed with 70% ethanol, dried and dissolved in water.

Amplification of DNA with PCR

All primer sequences for neutral markers were same as in Dettman *et al.* (2003). Primers for *frequency* and *white collar-1* genes were designed using *frq* (NCU02265.4) and *wc-1* sequence (NCU2356.4) respectively, deposited at Broad Institute (http://www.broadinstitute.org/annotation/genome/neurospora/MultiHome.html). The *frq* promoter sequence (chromosome IV, supercontig 7, position 1123159-1125203) was obtained from the plasmid VG110 *frq* oLucI in pBM61 generously provided by J. C. Dunlap (Dartmouth Medical School, Hanover, NH), and corresponding primers were designed (Table 2.2).

The genes and promoter were PCR-amplified from genomic DNA with the following reaction conditions: 200 µM dNTPs (Qiagen, Hilden, Germany), 0.5 µM of each primer (Metabion, Planegg/Martinsried, Germany, Table 2.2), 0.02 U/µl Phusion DNA Polymerase (100 U; FINNZYMES, Espoo, Finland). The thermal cycler protocol was as follows: initial denaturation at 95°C for 2 min, 30 cycles of 95°C for 1 min, annealing temperature with steps down from 63°C to 54°C for 30 sec, 72°C for 3 min; 5 cycles of 95°C for 1 min, annealing at 54°C for 30 sec and 72°C extension for 3 min; 5 min final extension at 72°C; hold at 4°C.

For neutral markers the following PCR reaction conditions were used: 10 mM dNTPs (Qiagen, Hilden, Germany), 5 pmol/µl of each primer (Metabion, Planegg/Martinsried, Germany, Table 2.2), 10X Qiagen PCR Buffer, 5X Qiagen Q-Solution, 25 mM MgCl$_2$, 5 U/ml Taq DNA Polymerase (Qiagen, Hilden, Germany). The thermal cycler protocol was as follows: initial denaturation at 94°C for 2 min, 35 cycles of 94°C for 1 min, marker-specific annealing temperature for 30 sec,

72°C for 1 min; 8 min final extension at 72°C; hold at 4°C. Finally, amplification products were purified from the gel using QIAquick Gel Extraction Kit (Qiagen, Hilden, Germany) according to manufactures protocol and then used for sequencing.

Table 2.2: List of primers used in this thesis.

Locus	Primer (5' - 3')	Annealing temp. (°C)	Used for*
Frq gene	TTCGAGAACACCGGTACCTGA	61	PCR - rev
	ATAGTCTCAGGCTTCGAGGGC	63	PCR - fwd
	CGCCTTGCGCGAGATACTAG	61	Seq - fwd
	ACCGATGGAACCAAGTTCAG	58	Seq - fwd
	AAGCTTGCATCCAACTACAGGAT	61	Seq - fwd
	ACTACCGCAGTGTCATTGACGAC	57	Seq - fwd
	CTCTAACGAGGAATCCAGGTATCCG	59	Seq - fwd
	ACTGGGACTGGAAGCGGAGATGGAA	61	Seq - fwd
	ATAAGAATGGTCGGAGGAAGAAG	59	Seq - fwd
	TCCCAGTGCGGAAGATGAAG	59	Seq - rev
	ATGAAAGGTGTCCGAAGGTG	58	Seq - rev
	CGCCACCCGAGTTGGAT	57	Seq - rev
	TTGGATGCAGAACCATTGTCTT	62	Seq - rev
	TCACAAAATGGTCGTCAGGAAGTA	62	Seq - rev
	TCAGTCTCGGTATCGAACGCCAAATCG	61	Seq - rev
	GTCGTCCCAGCTGGAATCGGTACT	61	Seq - rev
	ACATCGGTTTGTAATGAAAGGTGT	60	Seq - rev
Frq promoter	GGACGTCGTTCAGTTTAATAACGC	64	PCR - rev
	GCAATGGAGTGTGTAAAGATTCAG	62	PCR - fwd
	CCGGCGACCATGCTGATTGATTGATTG	61	Seq - fwd
	GTCTCCTTTGATATGCCCAGAAAATCT	57	Seq - fwd
	CTGTCGAGAGCGAACTGTTGC	58	Seq - rev
	GTTGTCCGGCCAAACTCTGGAACCTGT	63	Seq - rev
	TGGAAAGTCCAAAAGCGCAATTGCGAG	60	Seq - rev
	CCTCATGTTTCTGCCAAAAAGCCATCGT	60	Seq - rev
	CAAACTTGTGTGTTCCCAAAATGCCTT	57	Seq - rev
Part I	GTGCTGGCATCCCTGTGTTGATA	65	Seq - fwd
	GCGGATATAACTTGTTAGCTCGAT	62	Seq - rev
Part II	CCAGGCTGTTGTGGAAAGTCACT	65	Seq - fwd
	CACTCTTTGGCAACTCTGAACCA	63	Seq - rev
White collar-1	GCCTTGATCTTGTAGTTGTTGCT	61	PCR - rev
Part I	TTTCCCGTCTGCTTGAGTGAC	61	PCR and seq - fwd
	GATGATTTCATCAAGAGAGTCGCT	62	Seq - rev
Part II	TATAGCTACTTCAGCCAATTCTGC	62	Seq - fwd
	ACCGATGAGTCATAAGAGGTCGA	63	Seq - rev

*Seq = sequencing, rev = reverse, fwd = forward.

DNA sequencing

Sequencing setup

For DNA sequencing, the Sanger method was used. Sequencing reactions were performed using Big Dye Terminator BDu3 (Applied Biosystems, Darmstadt, Germany) and the following conditions: 1.3 µl of 1 M sequencing primers (Metabion, Planegg/Martinsried, Germany, Table 2.2), 2 µl of 5X Sequencing Buffer (Applied Biosystems, Darmstadt, Germany), 1 µl of Big Dye (Applied Biosystems, Darmstadt, Germany), 2.7 µl of amplified product, and 3 µl of water. The PCR program was as follows: initial denaturation at 96°C for 1 min, 25 cycles of 96°C for 10 sec, annealing at 50°C for 5 sec, 60°C for 4 min; hold at 4°C. PCR products were purified from dye terminator nucleotides, primers, excess salts, and other contaminants from sequencing reactions with Sephadex G-50 (Sigma-Aldrich, Steinheim, Germany). The plate was filled with Sephadex and 300 µl H_2O and left for 2 h at room temperature. The excess water was removed by centrifuging on 6000 rpm for 4 min. The columns were washed with water and sequencing reactions were added. After centrifuging the plate on 6000 rpm for 4 min the cleaned sequencing reactions are collected. Formamide was added to reactions and left on 95°C for 5 min. Sequencing reactions were run on a DNA Sequencer 3100 (Applied Biosystems, Foster City, CA, USA).

For *wc-1* gene the sequencing reaction was prepared as follows: 1.3 µl of 1 M sequencing primers (Metabion, Planegg/Martinsried, Germany, Table 2.2), 2.7 µl of amplified product, and 3 µl of buffer EB (Qiagen, Hilden, Germany). The mixture was send to Sequencing Service at LMU (http://www.gi.bio.lmu.de/sequencing).

Cloning of the PCR products was not necessary because *Neurospora* strains are haploid and possessed only one allele per individual. Sequence data were examined and edited visually using Sequencher 4.7 (Gene Codes, Ann Arbor, MI, USA). Nucleotide sequences from neutral markers have been deposited in GenBank under the accession numbers JQ629968–JQ630031.

Phylogenetic analysis

DNA sequences were aligned using Clustal W (http://www.ebi.ac.uk/Tools/msa/clustalw2/) and edited manually with MacClade 4.06 (Maddison and Maddison, 2000). For all alignments, the gaps were treated as missing data.

For analysing neutral markers, the regions of sequence with ambiguous alignment and microsatellite repeats were excluded as in Dettman *et al.* (2003). Sequence data from TMI, DMG, TML and QMA loci were aligned in one file and added to the alignment file provided by Villalta *et al.* (2009). However, to simplify the analysis, a total of 67 *Neurospora* isolates contributed to the present phylogenetic analysis (see Figure 3.2).

The appropriate nucleotide substitution model was chosen using Modeltest 3.7 (Posada and

Crandall, 1998) and PAUP 4.0b10 (Swofford, 1998). The alignment and the chosen nucleotide substitution model were used as input for MrBayes (Huelsenbeck et al., 2001; Ronquist and Huelsenbeck, 2003). The analysis was run for 1 million generations with burn-in of 2 500 generations to produce a consensus tree with Bayesian posterior probabilities. The final trees presented in figures, and bootstrap branch support values were obtained with RAxML v7.2.8 (Stamatakis, 2006; Stamatakis et al., 2008) using the maximum likelihood option with gamma model for 100 replicates.

The substitutions per site were calculated with PAUP 4.0b10 using maximum likelihood criteria as branch length from the corresponding isolate to the root of the tree. Kimura two-parameter overall mean genetic distance was calculated with MEGA 5.05 (Tamura et al., 2011).

Physiological methods

Growth conditions

Light-dark cycles

All light-dark cycles were performed at a constant temperature of 25°C and were carried out in light-tight boxes which contained an air-circulating fan, a white fluorescent strip lamp (3 $\mu E/m^2/sec$, 10 W, OSRAML) and a layer of diffuser (Cinegel #3026, Rosco) to improve light distribution (Figure 2.1). The cycle length (T) was 24 h and 13 different photoperiods were applied from complete darkness (DD) to complete light (LL) in steps of 2 h. Each experiment was continuously monitored for temperature and light using Data Loggers (HOBOwareLite 2.6.0, Bourne, MA).

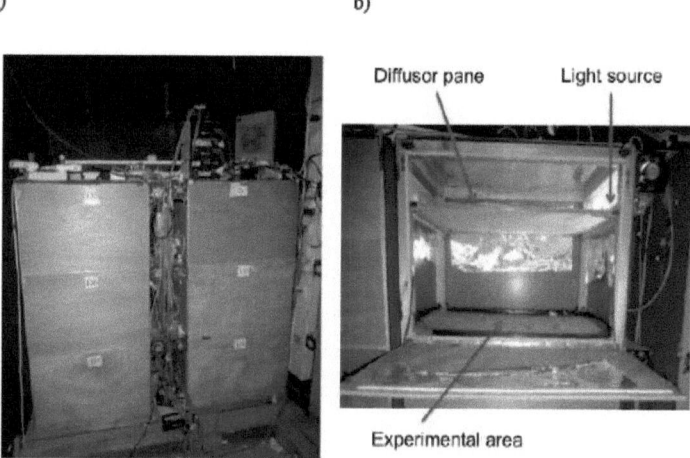

Figure 2.1: a) Six light boxes used to perform light-dark cycles; b) the interior of the light box.

Temperature cycles

Temperature cycles were created in custom-made incubators (Figure 2.2). Two water-baths alternately circulated warm or cold water through metal tubes at the bottom of the boxes to achieve the proper temperature. Large enclosed water baths ensured 100% humidity and smooth, gradual and slow temperature transitions.

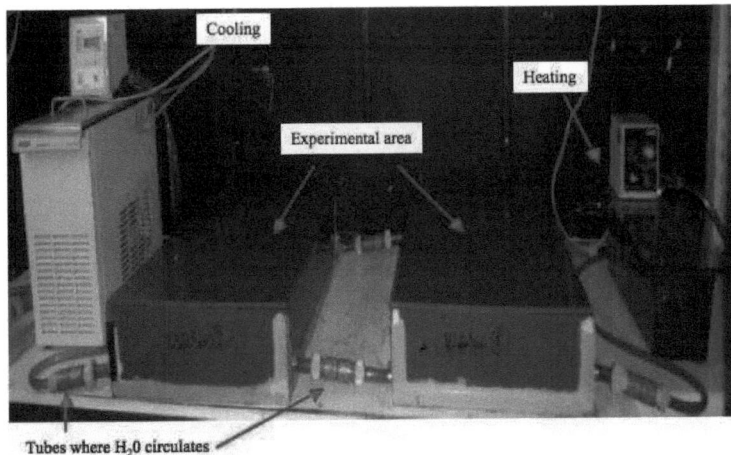

Figure 2.2: Custom-made incubators for performing temperature cycles.

All temperature experiments were performed with a cycle length of 24 h in DD (total darkness). Temperature pulses were delivered as step up from 22°C (cold) to 27°C (warm) for different time periods (16, 25, 33, 40, 50, 60, 67, 75, 84 % of warm of the cycle length) followed by a step down to 22°C. Each experiment was continuously monitored for temperature and light using Data Loggers (HOBOwareLite 2.6.0, Bourne, MA).

Race tube assay

Race tube setup

Race tubes are hollow glass tubes (40 cm long, 12 mm diameter; Höhn, Munich, and Schmitz, Munich, Germany) whose both ends are bent up 45° (Figure 2.3a). They were used to monitor the growth of the isolates and to measure the period and the phase of the circadian conidiation.

Packs of six race tubes (five for the isolate investigated and one for the control strain) were used and each race tube was filled with 8 ml of molten Race Tube Media (Appendix). Both ends of the tubes were closed with cotton plugs. After autoclaving, the race tubes are swirled to remove condensed water drops, which could eventually drip onto the agar surface and interfere with the continuous mycelial growth. Placed on an even surface, the agar was left overnight to cool and solidify. Using a loop, the race tubes were then inoculated at one end with conidia from slant cultures (see above). The tubes were placed at room temperature overnight in constant light to germinate and then transferred to the desired experimental conditions (starting with lights off or the colder temperature). The growth front of the mycelia on the agar was marked with a pen before the transfer and then every other day. When the mycelia reached the other end of the race tubes, the tubes were scanned (scanner: Microtek ScanMaker 9800XL; program settings: Greyscale; Image Resolution: 72 dpi) from underneath (Figure 2.3b) for analysis with the ChronOSX Program 2.1 (a new version of CHRONO Program, Roenneberg and Taylor, 2000).

a)

b)

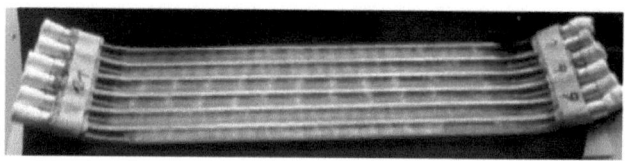

Figure 2.3: a) A six-pack race tubes; b) A scanned race tube, the conidial bands and marks are evident.

Data analysis

Since the growth rate of *Neurospora* is more or less constant, both period length (time after which a definite marker of the oscillation reoccurs) and phase angles (time given in hours or degrees at which conidiation onset occurs, relative to the *zeitgeber*) of the circadian rhythm of conidiation can be calculated from the position of bands relative to one another and to the growth fronts that were marked daily.

For this, the protocols with date and time of each mark were saved as text-only files and together with the race tube scans imported into the ChronOSX Program and used for further analysis. The ChronOSX Program allowed the production of graphs showing the rhythm of conidiation, plotted as relative pixel density versus time/day of an experiment. Conidiation was quantified by the number of white pixels in each vertical line of the image and expressed as deviation around the non-rhythmic trend and graphed as double plots (two days of an experiment arranged in a row) in order to visualize the rhythm across midnight (Figure 2.4).

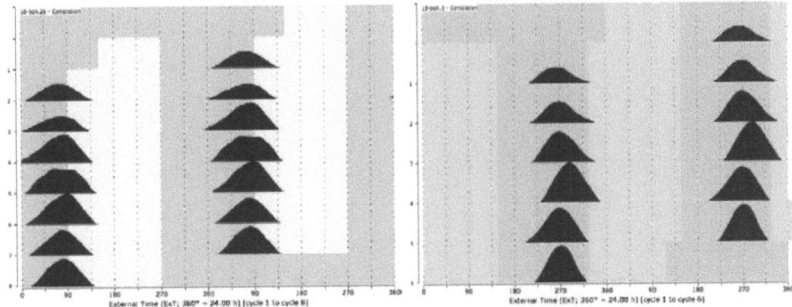

Figure 2.4: Example of double plots. *Zeitgebers* are shown as yellow and red bars (light and warm temperature, respectively) compared to grey and blue (darkness and colder temperature, respectively). The conidiation bands are shown in dark blue.

Omitting the first and the last day of each recording, daily profiles of *Neurospora*'s banding rhythm were calculated from the data of density scans of single race tubes, with a smoothing factor of 4 h.

The period was calculated as periodogram in ChronOSX in hours. The notation FRP was used for the free-running period in constant darkness, while symbol τ_E for the period in entrained conditions (light-dark or temperature cycles, afterward called "entrained period"). The isolates were defined as arrhythmic where the peak of conidiation was under the p-value line (p<0.01).

For phase analysis, a two-component cosine function was fitted to each profile and a trend corrected for 24 h. Minimum, onsets (upward transition through the zero line of the cosine fit), maximum, and offsets (downward transition through the zero line of the cosine fit) were calculated based on the two-harmonic cosine curve. Phase reference points were plotted in degrees (360° per cycle) and calculated to external time. Phase was assessed as onset of conidiation relative to mid-dark/cold (Phi on) for each isolate, and only those race tubes were included in the analysis, which were stably entrained ($\tau_E = T \pm 1h$).

Statistical analysis

All data were analysed using the statistical Software Prism 4.0 (Graphpad Software, Inc., La Jolla, CA). All data were tested for normality with Shapiro-Wilk test. Unpaired t-tests and one-way

ANOVAs followed by post-hoc Bonferroni's or Dunn's Multiple Comparison Tests were used to test for differences between the groups. A linear regression analysis was used to find correlations (Pearson r or Spearman r). The cluster analysis was done with SPSS Statistics 19 (Inc., Chicago IL), using Ward's method of hierarchical clustering and similarity as squared Euclidian distance, data were standardized with z score by variables.

3. RESULTS

Genetic analysis

The aim of this part was to test if there are differences in DNA sequences of clock genes in different *N. crassa* isolates and if these differences correlate with latitude. Since Michael *et al.* (2007) found that simple sequence repeats (SSRs) in *white collar-1* gene correlate with the latitude of the origin and are responsible for the local adaptation, these repeats were studied here in other set of isolates. However, Michael *et al.* (2007) did not consider the phylogenetic groups, so they were added to analysis in this thesis. Furthermore, one more clock gene, *frequency*, was analysed here.

About 109 000 nucleotides of new sequences were obtained. These were used to see the genetic variability of the isolates in four neutral markers, *frequency* and *white collar-1* gene. Furthermore, the existence of the latitudinal cline in DNA sequences was tested. The isolates coming from latitudes between 0° and 30°N were consider as "test group", as these isolates are found in all publications presented to date.

Neutral markers

In order to see if the genetic differences presented in genes or regulatory elements are due to phylogenetic species groups, four neutral markers were sequenced. These markers (TMI, DMG, TML and QMA) represent four independent nuclear loci found at VR, IVR, VL and IIR chromosome/linkage group, respectively. TML codes for trimethyllysine dioxygenase, TMI for hypothetical protein, which existence has not been verified yet, and other two (DMG and QMA) have not known function. Dettman *et al.* (2003) used these markers for phylogenetic species recognition in *Neurospora* and clearly showed that the specie *Neurospora crassa* has three subgroups: NcA, NcB, and NcC.

Approximately 32 000 nucleotides of new sequence data are reported in this thesis. Summaries of the alignments for the four loci are shown in Table 3.1. The loci differed in the amount and form of variation; all of them presented microsatellite repeats, single nucleotide polymorphisms (SNPs) and insertions/deletions (indels) in the sequence. For example, QMA was the most variable locus and DMG the least, where the average genetic distance was 0.0077 and 0.0213 for DMG and QMA, respectively. The QMA locus had fourfold number of parsimony informative characters than DMG locus.

Table 3.1: Summary of the six DNA sequence alignments.

	TMI	DMG	TML	QMA	Combined*	Combined with Villalta et al. (2009)
Number of isolates†	26 (10)	26 (10)	26 (10)	17 (1)	26 (10)	67 (51)
Length of final alignment	446	410	611	510	2014	2148
Excluded positions due to ambiguous alignment	none	374-389	140-147; 376-401	none	820-835; 998-1005; 1233-1258	846-861; 1027-1035; 1274-1298
Position of omitted microsatellites	383-429	229-262	after base 579	434-502; 448-486	after bases 383, 675, 1438, 1901, and 1915	after bases 382, 700, 1477, 1947, and 1965
Number of constant characters	413	385	537	434	1766	1528
Number of parsimony informative characters	24	8	26	34	117	396
Average genetic distance, d	0.0161	0.0077	0.0173	0.0213	0.0146	0.0272

* sequences from four loci were concatenated and analyzed as one sequence.
† number of isolates taken from Dettman *et al.* (2003), Jacobson *et al.* (2006) and Villalta *et al.* (2009) are given in brackets.

As shown in Figure 3.1, there were some well-supported clades common to the single-locus trees. For example, clade B isolates (in blue) formed one well-supported group in TMI, DMG and QMA loci with Bayesian posterior probabilities (PP) of 1.00, and maximum likelihood bootstrap proportions (MLBP) of 100%, 96% and 99%, respectively. One clade C isolate (in green) was basal to all other clades in QMA and TML loci trees. Pakistan and Thailand isolates (in orange) formed well-supported group, basal to other clades, only in TMI locus tree with Bayesian posterior probabilities (PP) of 1.00, and maximum likelihood bootstrap proportions (MLBP) of 100%.

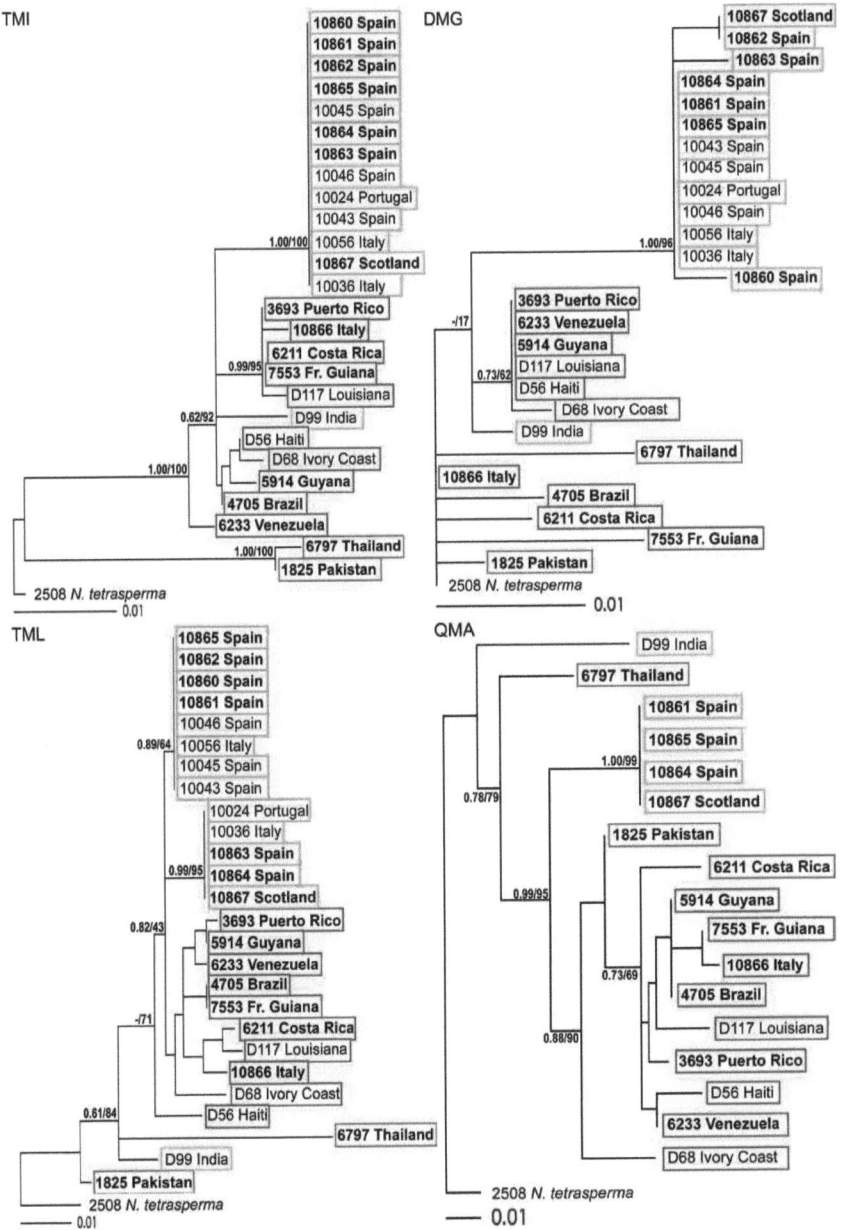

Figure 3.1: Maximum likelihood trees produced from each of the four single locus data. *N. tetrasperma* was used as outgroup. Numbers near branches indicate confidence levels (Bayesian posterior probability/maximum likelihood bootstrap proportion); newly characterized isolates (see Table 2.1) are indicated in bold; numbers in front of collection sites indicate either the FGSC number or were taken from Dettman *et al.* (2003; prefix D); the sequence of the isolates in normal font were from Jacobson *et al.*, 2006. Bar indicates substitutions per site.

More *Neurospora* species where included into analysis of final phylogenetic tree to test the relationship between newly added isolates and other *Neurospora* species. The phylogenetic analysis on combined dataset of all four neutral markers yielded one maximum likelihood tree (Figure 3.2). The tree had Bayesian posterior probabilities (PP) between 0.51 and 1.00, and maximum likelihood bootstrap proportions (MLBP) between 15% and 100%. It confirmed the phylogenetic species of *Neurospora* and was similar to those provided by Dettman *et al.* (2003) and Villalta *et al.* (2009).

Neurospora crassa was grouped in three known clades (A, B and C) and their geographical distribution was also confirmed (Dettman *et al.*, 2003). The nine isolates from the Caribbean Basin (Louisiana, Puerto Rico, Haiti, Costa Rica, Ivory Coast, Guyana, Venezuela, Fr. Guiana, and Brazil) and one European isolate (10866 Italy) belonged to clade A (in red); the other 13 European isolates (from Scotland, Italy, Spain and Portugal) to clade B (in blue). The branch support values for clades A and B were equally high (Bayesian PP/MLBP = 0.93/50% and 1.00/100%, respectively). For Indian isolates (clade C, in green), the values were 1.00/75% compared to 1.00/95% of clade A and B combined (Figure 3.2). Two isolates (from Pakistan and Thailand) did not fall into any of the existing *N. crassa* clades and were found to be separate and basal to all other clades with Bayesian PP/MLBP of 0.87/60%.

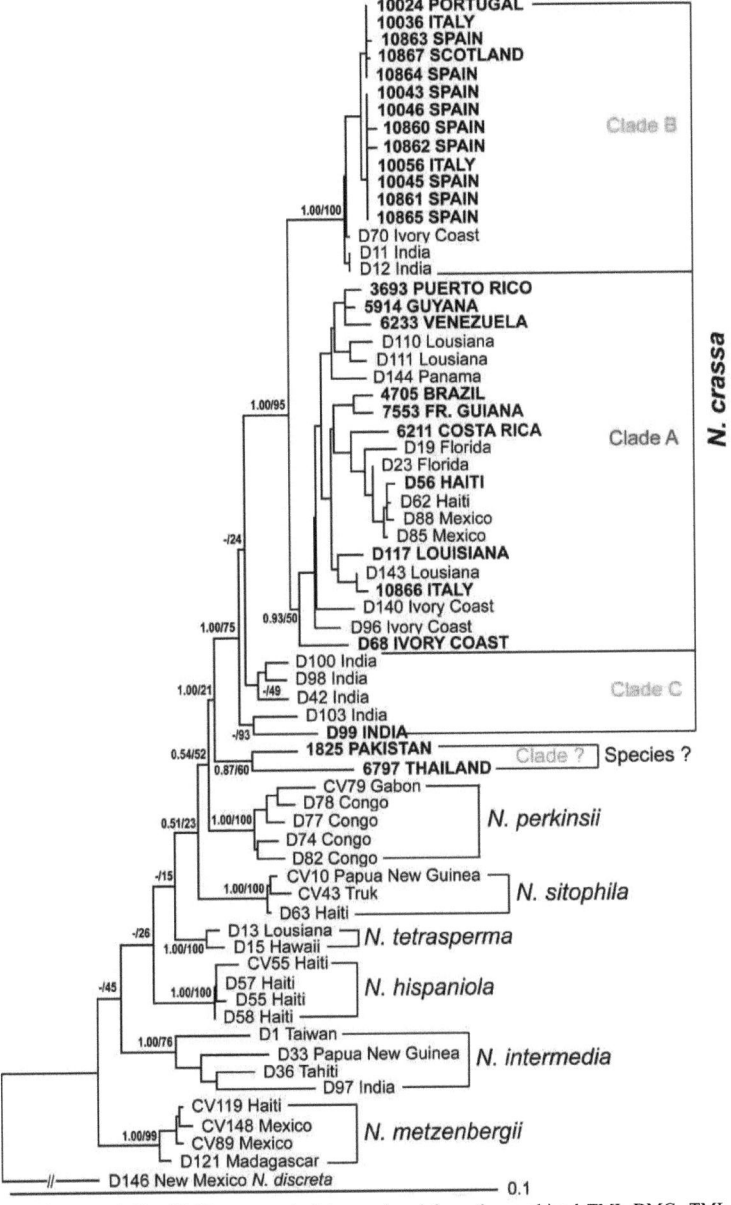

Figure 3.2: Phylogram of 67 wild *Neurospora* isolates produced from the combined TMI, DMG, TML, and QMA sequences. *N. discreta* was used as outgroup. Different colours depict different clades (Dettman *et al.*, 2003). Numbers near branches indicate confidence levels (Bayesian posterior probability/maximum likelihood bootstrap proportion); isolates used in this study (see Table 2.1) are indicated by capitals in bold; numbers in front of collection sites indicate either the FGSC number or were taken from Dettman *et al.* (2003a; prefix D) or from Villalta *et al.* (2009; the prefix CV). Bar indicates substitutions per site.

Figure 3.3 shows more in detail the phylogenetic tree of the 26 isolates used in this thesis, which are divided into three clades. Here, the branch support values for clades A and B were: Bayesian PP of 0.98 and 1.00, and MLBP of 72% and 100%, respectively. For Indian isolate, the values were 0.99/83% compared to 1.00/95% of clade A and B combined (Figure 3.3). Two isolates Pakistan and Thailand formed a separate group with well-supported Bayesian PP/MLBP of 1.00/92%.

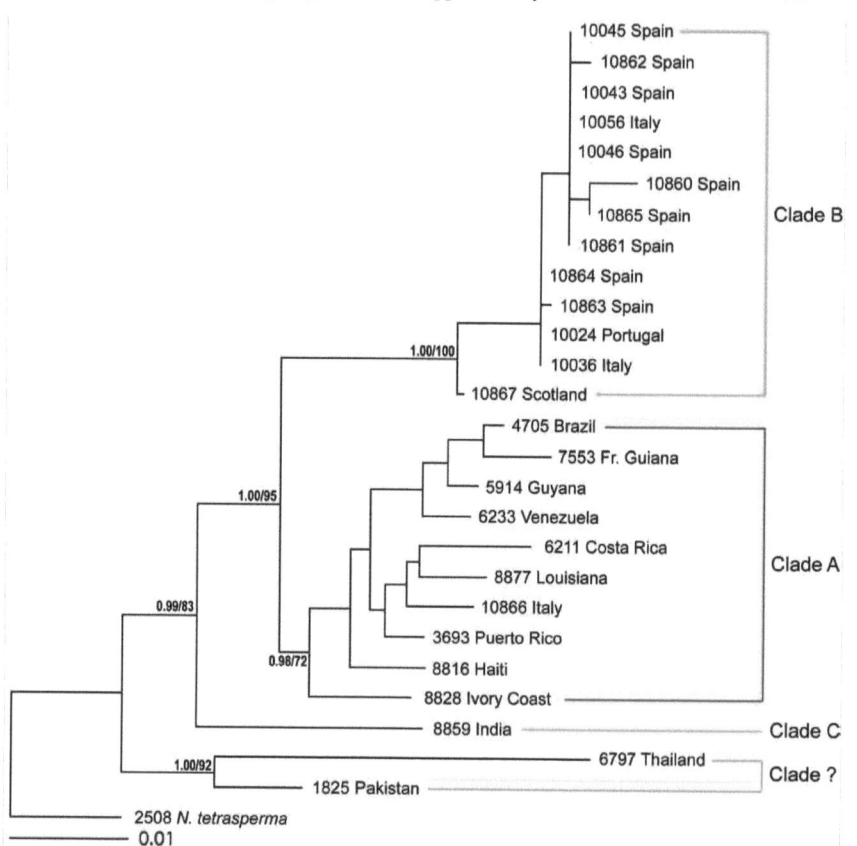

Figure 3.3: Maximum likelihood phylogram produced from TMI, DMG, TML and QMA loci combined of 26 *Neurospora crassa* isolates used in this thesis. *N. tetrasperma* was used as outgroup. The different colours depict different clades. Numbers near branches indicate confidence levels (Bayesian posterior probability/maximum likelihood bootstrap proportion); numbers in front of collection sites indicate the FGSC number. Bar indicates substitutions per site.

Substitutions per site of four neutral markers combined for all 26 isolates did not correlate with latitude of the origin ($p = 0.1821$, $r = 0.2701$) or in each clade separately (clade A: $p = 0.4858$, $r = -0.2501$; clade B: $p = 0.3458$, $r_S = -0.2847$; data not shown). A positive correlation between substitutions per site of each neutral marker separately and latitude of the origin was found in TMI ($p < 0.0001$, $r_S = 0.8468$) and DMG ($p = 0.0062$, $r_S = 0.5220$) loci but not in TML ($p = 0.3641$, $r_S =$

-0.1856) and QMA (p = 0.2719, r = -0.2825) loci (Figure 3.4). If only the isolates from "test group" were considered, the correlation was not more significant in all loci (TMI: p = 0.1418, r = 0.4730; DMG: p = 0.2141, r_S = -0.4110; TML: p = 0.0656, r_S = 0.5780; QMA: p = 0.8664, r = 0.0576; data not shown).

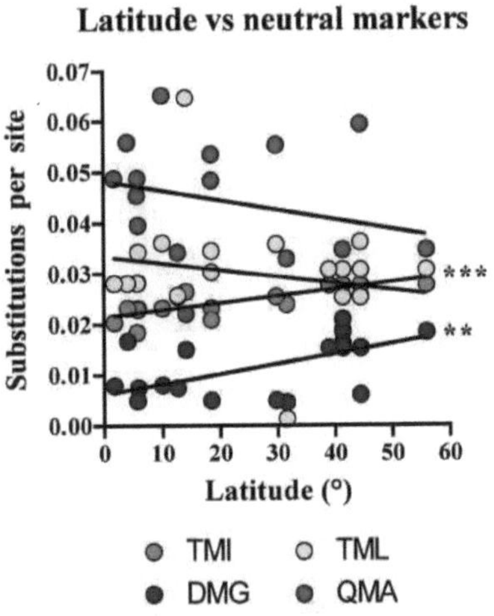

Latitude vs neutral markers

○ TMI   ○ TML
● DMG   ● QMA

Figure 3.4: Linear correlation between latitude and substitutions per site in four neutral markers.

Since a number of Spanish isolates were genetic identical, they were considered as one individual (e.g., in TMI locus all Spanish isolates, in DMG locus Spain 10043, 10045, 10046, 10861, 10864, 10865; in TML locus Spain 10043, 10045, 10046, 10860 – 10862, 10865, and 10863 – 10864; in QMA locus Spain 10861, 10864, 10865). Considering single loci and dividing the isolates into clades, no correlation was found in any locus (DMG: clade A: p = 0.2325, r_S = -0.4202; clade B: p = 0.9235, r = 0.0408; TML: clade B: p = 0.8028, r_S = 0.1050; QMA: clade A: p = 0.2056, r = 0.4379; Figure 3.5). Exception was clade A, which showed positive correlation in TMI (p = 0.0397, r = 0.6553) and TML (p = 0.0327, r = 0.6737) loci (Figure 3.5).

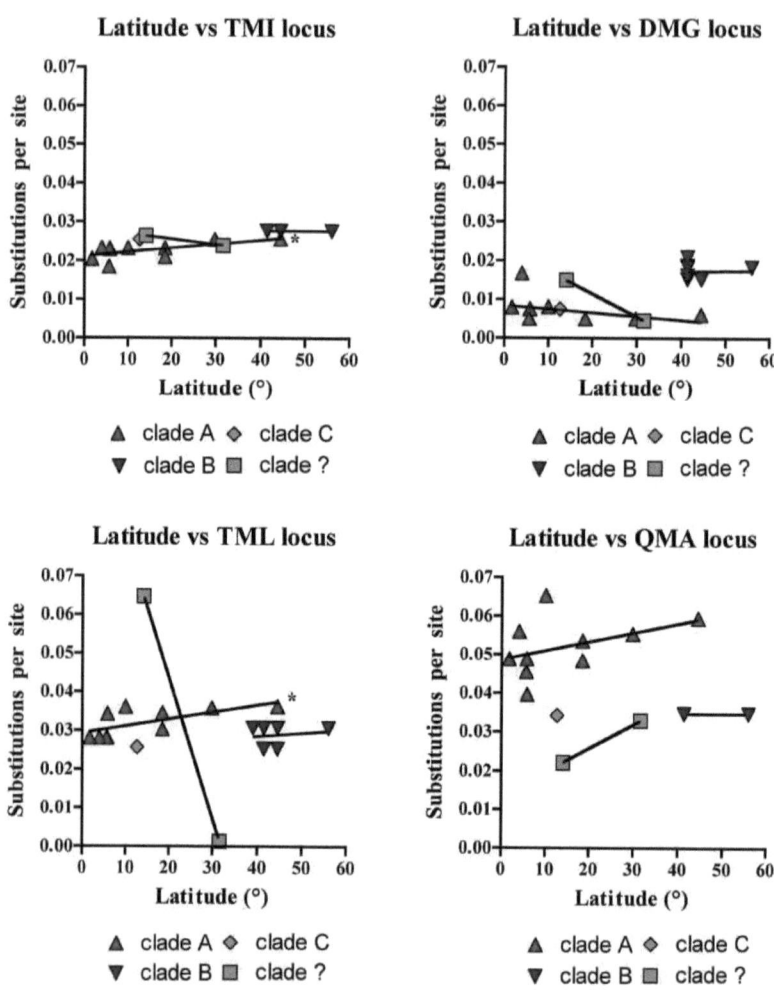

Figure 3.5: Correlation between latitude and substitutions per site in each of the four neutral markers.

Since, Pakistan and Thailand isolates were identified as members of a new group or even of new *Neurospora* species, they were excluded from further analysis.

White collar-1 gene

Two regions of *white collar-1* genomic sequence containing simple sequence repeats (SSRs) were sequenced (part I containing 5'AG/GA and NpolyQ repeats; part II containing CpolyQH repeat). More than 37 000 nucleotides of new sequence were obtained. Since isolates from Spain (10860 – 10864, 10043, 10045) were identical in phylogenetic analysis (see Figure 3.7), they were analysed as one individual. The amount of repeats across the whole population of isolates showed that NpolyQ repeat was the longest, while CpolyQH was the shortest one (Table 3.2). 39% of the isolates had 5'AG/GA 14 bases long, while 39% had NpolyQ 36 amino acids long and 39% CpolyQH only of 9 amino acids.

Table 3.2: The percentage of isolates and amount of repeats.

N of repeats	5' AG/GA	NpolyQ	CpolyQH
44		5.56%	
42		5.56%	
38		5.56%	
36		27.78%	
35		38.89%	
26		16.67%	
18	27.78%		
15	5.56%		
14	38.89%		
13	22.22%		
10			11.11%
9	5.56%		38.89%
8			27.78%
6			5.56%
5			16.67%

No correlation between latitude and number of repeats in *white collar-1* gene was found (5'AG/GA: p = 0.1384, r_S = 0.3747; NpolyQ: p = 0.8615, r_S = -0.0458; CpolyQH: p = 0.8076, r_S = -0.0639; Figure 3.6). The correlation was also not evident considering the "test group" (5'AG/GA: p = 0.7330, r_S = 0.1268; NpolyQ: p = 1.0000, r_S = 0.0074, CpolyQH: p = 0.7589, r_S = -0.1172; data not shown). Furthermore, no correlation between latitude and number of repeats in *wc-1* could be found separating isolates into clades (clade A: 5'AG/GA: p = 0.0883, r_S = 0.5742; NpolyQ: p = 0.7330, r_S = 0.1268; CpolyQH: p = 0.8113, r_S = 0.0921; clade B: p = 0.7825, r_S = -0.1498; data not shown).

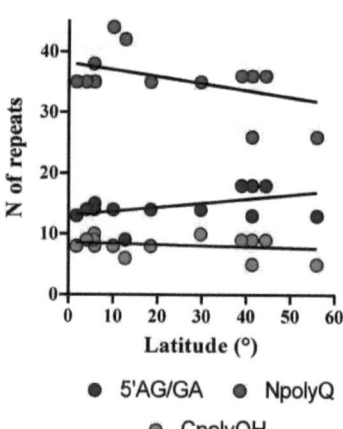

Figure 3.6: Correlation between latitude of origin and number of repeats in *wc-1*.

The phylogenetic analysis on the combined dataset of two regions of *wc-1* gene yielded maximum likelihood tree (Figure 3.7). Four groups based on well-supported Bayesian PP and MLBP were identified: Scotland and Spanish isolates formed one group (Bayesian PP/MLBP = 1.00/100%); the second group comprised isolates 10036 Italy, 10056 Italy, 10866 Italy, 10046 Spain, 10024 Portugal, 6233 Venezuela and 8877 Louisiana (Bayesian PP/MLBP = 0.99/72%); the third group Caribbean Basin isolates (Bayesian PP/MLBP = 0.99/98%), and Indian isolate formed one separate group. The differences found at *wc-1* locus are not due to clade properties of the isolates.

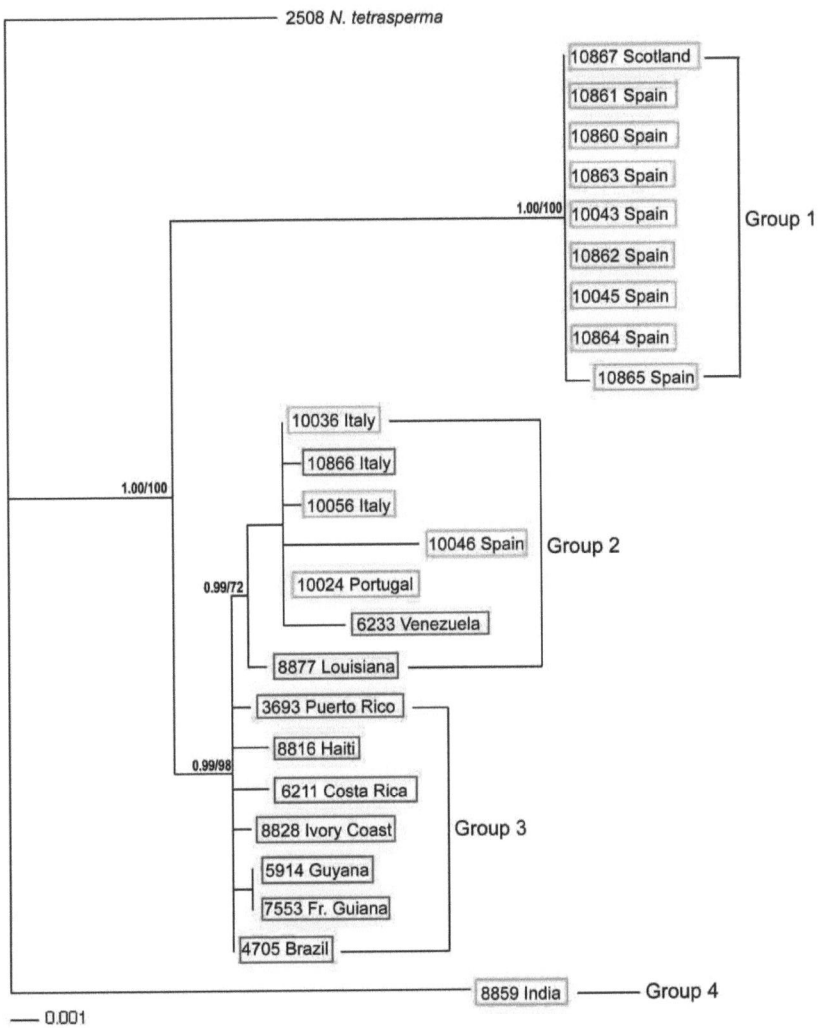

Figure 3.7: The maximum-likelihood tree of *white collar-1* locus for 24 isolates. Different colours depict different clades (red = clade A, blue = clade B, green = clade C). *N. tetrasperma* was taken as outgroup. Numbers near branches indicate confidence levels (Bayesian posterior probability/maximum likelihood bootstrap proportion); numbers in front of collection sites indicate the FGSC number. Bar indicates substitutions per site.

Only group 2 showed correlation between latitude and number of repeats in 5'AG/GA ($p = 0.0341$, $r_S = 0.7977$) when the isolates were separated into groups found with phylogenetic analysis of *wc-1*. No correlation was found in group 2 isolates in other SSRs (NpolyQ: $p = 0.7825$, $r_S = -0.1348$; CpolyQH: $p = 0.3536$, $r_S = -0.4119$). Group 3 isolates showed no correlation in any of SSRs

(5'AG/GA: p = 0.3956, r_S = 0.4009; NpolyQ: p = 0.6615, r_S = 0.2041: CpolyQH: p = 0.3024, r_S = -0.4454; data not shown).

Positive correlation between latitude of the origin and the number of substitutions per site in *white collar-1* considering all isolates was found (p = 0.0060, r_S = 0.6207; Figure 3.8). However, the correlation was not significant testing the isolates from the "test group" (p = 0.4697, r_S = 0.2622) or separating isolates into clades (clade A: p = 0.3129, r_S = 0.3537; clade B: p = 0.5000, r = 0.3091; data not shown) or phylogenetic groups of *wc-1* (group 2: p = 0.9635, r_S = 0.0180; group 3: p = 0.1629, r = 0.5903; Figure 3.8).

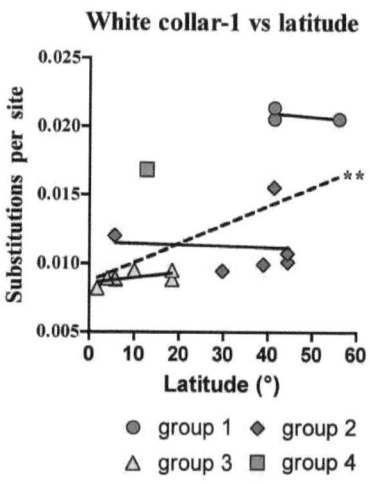

Figure 3.8: Correlation between latitude and substitutions per site in phylogenetic groups of *white collar-1* gene. Dashed liner regression line refers to all isolates.

The part I of *wc-1* had more substitutions per site than part II in all isolates (maximum of 0.0178 in 10860 – 10884 Spain, 10043 and 10045 Spain and minimum of 0.0012 in 10046 Spain; data not shown). On average there were 0.012 substitutions per site difference between two parts. The number of substitutions in part I correlated with those in part II (p = 0.0271, r_S = 0.5195), meaning the more substitutions per site in part I the more in part II (Figure 3.9). If the isolates were divided into "test group" (p = 0.4918, r_S = 0.2425; data not shown), clades (clade A: p = 0.7330, r_S = -0.1185; clade B: p = 0.0953, r_S = 0.4820; data not shown) or phylogenetic groups of *wc-1* gene (group 2: p = ns, r_S = -0.0619; group 3: p = ns, r_S = -0.4714; Figure 3.9) the correlation was not significant.

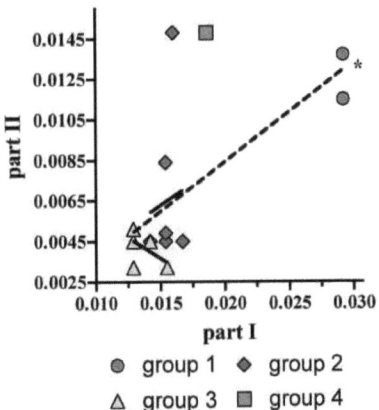

Figure 3.9: Correlation between substitutions per site in part I and II of the *white collar-1* gene. Different groups correspond to the phylogenetic groups of *wc-1*. Dashed linear regression line refers to all isolates.

Frequency locus

The coding region of the clock gene *frequency* and its promoter in three isolates (10867 Scotland, 10861 Spain, and 8877 Louisiana) were completely sequenced. For the coding region, the 10861 Spain isolate showed the most variation compared to the outgroup strain *N. tetrasperma*, with similarity of 73% between them (Table 3.3). Numerous single nucleotide polymorphisms (SNPs) but no insertions/deletions (indels) for the coding region were found. The promoter, however, showed many indels, which ranged from one to 20 nucleotides in length. The most variable isolate, compared to *N. tetrasperma*, originated from Louisiana (59% of similarity) and the least variable from Spain (76% of similarity; Table 3.3). On average, the *frequency* gene had a similarity of 89%, while its promoter of 67%. Of 2970 total characters in *frq* gene 2834 were constant, 111 were variable and 25 parsimony informative, while for *frq* promoter from 2964 characters 2747 were constant, 165 variable and 52 parsimony informative. This result showed that *frq* promoter was more variable than *frq* gene.

Table 3.3: Similarity between three isolates in *frequency* gene (above diagonal) and promoter (below diagonal). Numbers in front of the collection site indicate Fungal Genetic Stock Center numbers.

	10867 Scotland	10861 Spain	8877 Louisiana	*N. tetrasperma*
10867 Scotland	-	75.3%	96.7%	97.2%
10861 Spain	69.3%	-	72.8%	73.3%
8877 Louisiana	76.8%	58.5%	-	97.0%
N. tetrasperma	61.2%	76.4%	59.1%	-

The phylogenetic analysis gave one maximum likelihood tree, which had same topology for *frq* gene and promoter. The 10867 Scotland and 10861 Spain isolates formed one sister-group, while 8877 Louisiana isolate formed other group basal to the first one (data not shown). *Frq* promoter had more substitutions per site than *frq* gene for each isolate (Figure 3.10). However, the number of substitutions did not correlate with latitude of the origin, and the number of substitutions in *frq* promoter did not correlate with number of substitutions in *frq* gene (data not shown).

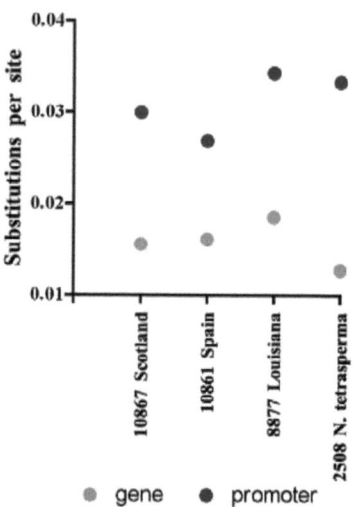

Figure 3.10: Number of substitutions per site in *frq* gene (orange dots) and *frq* promoter (blue dots) for three isolates. The isolates are arranged according their latitude. The numbers in front of the collection sites indicate FGSC numbers.

Frequency promoter

To take advantage of numerous informative characters (twofold in promoter than in gene) and indels present in the data, the analysis was continued only with the promoter. Two hot-spot regions (part I and II) were sequenced in 24 isolates, and a total of about 32 000 nucleotides were obtained.

Phylogenetic analysis under maximum likelihood was performed on the *frequency* promoter (part I + part II) alignment (Figure 3.11). The isolates were classified into three groups based on well-supported Bayesian PP and MLBP: the first group comprised Spanish and Scotland isolates (Bayesian PP/MLBP = 1.00/76, the second group comprised Italian isolates, 10046 Spain, 10024 Portugal, 8877 Louisiana, and 8859 India isolates (Bayesian PP/MLBP = 0.92/56), and the third group was composed of Caribbean Basin isolates (from Puerto Rico, Haiti, Brazil, Guyana, Fr. Guyana, Venezuela, Ivory Coast, and Costa Rica; Bayesian PP/MLBP = 1.00/100). Again like for *wc-1* gene, the clade properties are not responsible for the differences found in the *frq* promoter.

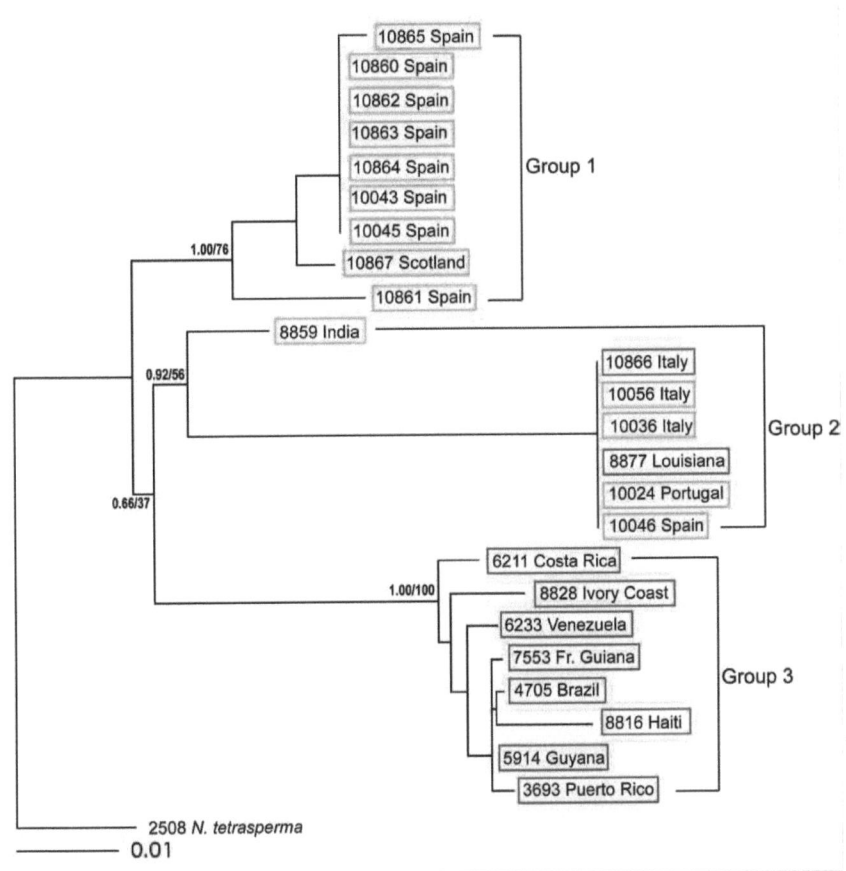

Figure 3.11: Maximum likelihood phylogram of *frequency* promoter for 24 *N. crassa* isolates. Different colours depict different clades (red = clade A, blue = clade B, green = clade C). *N. tetrasperma* was taken as outgroup. Numbers near branches indicate confidence levels (Bayesian posterior probability/maximum likelihood bootstrap proportion); numbers in front of collection sites indicate the FGSC number. Bar indicates substitutions per site.

Here, Spanish isolates (10860, 10862 – 10864, 10043, 10045) showed no differences in sequence and were considered as one individual. No correlation was found between latitude of the origin and substitutions per site in *frq* promoter considering all isolates together ($p = 0.9983$, $r = 0.0005$; Figure 3.12). When the isolates were separated according to the phylogenetic groups of *frq* promoter, none group showed correlation (group 1: $p = 0.5355$, $r = -0.4645$; group 2: $p =$ ns, $r = 0.6179$; group 3: $p = 0.1099$, $r = 0.6079$; Figure 3.12). The isolates from clade A, however, showed a positive correlation between substitutions per site and latitude of the origin ($p = 0.0028$, $r = 0.8326$; data not shown).

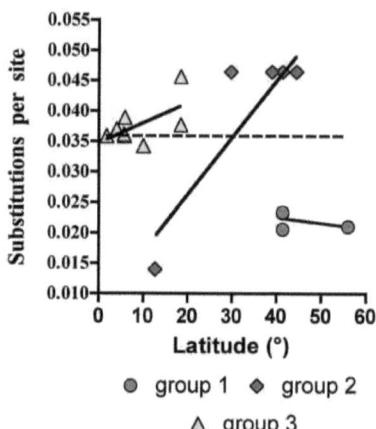

Figure 3.12: Correlation between latitude and substitutions per site in *frq* promoter. Different groups are derived from phylogenetic analysis of *frq* promoter. Dashed linear regression line refers to all isolates.

In 66% of the isolates part I had more substitutions per site than part II. On average, there were 0.023 substitutions per site difference between parts I and part II. There was no correlation between substitutions per site between part I and II in *frq* promoter when all isolates were considered (p = 0.1968, r_S = 0.2730; Figure 3.13). Furthermore, no correlation in the "test group" (p = 0.9184, r_S = 0.0408) was found, neither in the phylogenetic groups 1 and 3 of *frq* promoter (group 1: p = 0.1777, r_S = 0.5000; group 3: p = 0.7633, r = 0.1276; Figure 3.13) or clade A (p = 0.1231, r_S = -0.5207; data not shown). However, positive correlation was found in the phylogenetic group 2 (p = 0.0004, r_S = 1.000; Figure 3.13) and in clade B isolates (p = 0.0073, r_S = 0.7037; data not shown).

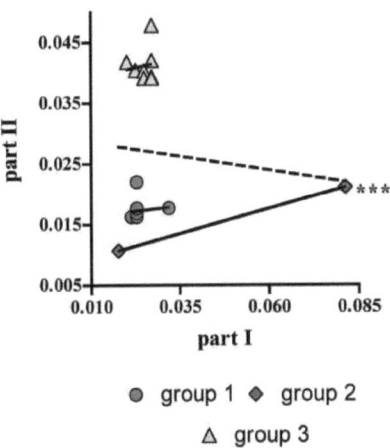

Figure 3.13: Correlation between substitutions per site in part I and II of the *frequency* promoter. Different groups correspond to the phylogenetic groups of *frq* promoter. Dashed linear regression line refers to all isolates.

Comparison between clock genes and clades phylogeny

To obtain the phylogenetic relationship of the clock genes, the maximum likelihood analysis was performed on combined data set from *wc-1* and *frq* promoter together. The analysis gave one maximum likelihood tree (Figure 3.14). There were four groups identified based on well-supported Bayesian PP and MLBP scores: group 1 comprised Spanish and Scotland isolates (with Bayesian PP/MLBP of 1.00/100), group 2 comprised Italian isolates, 10046 Spain, 10024 Portugal, and 8877 Louisiana isolates (Bayesian PP/MLBP = 1.00/100), most of the Caribbean Basin isolates formed group 3 (Bayesian PP/MLBP = 1.00/100), and basal was group 4, which comprised isolate from India. As for *wc-1* and *frq* promoter alone, the phylogeny based on neutral markers and clock genes do not share topologies, thus the clade properties are not responsible for the differences found in the clock genes. Furthermore, there were small differences between groups obtained with *white collar-1* gene (Figure 3.7) and *frq* promoter (Figure 3.11) alone and *wc-1* and *frq* promoter combined (Figure 3.14). In all three trees group 1 comprised Scotland and Spanish isolates. Group 2 of *wc-1* gene comprised Venezuela isolate, which was in the group 3 of *frq* promoter and combined analysis. India isolate was basal to all groups in *wc-1* and combined tree, while it was a part of group 2 in *frq* promoter tree.

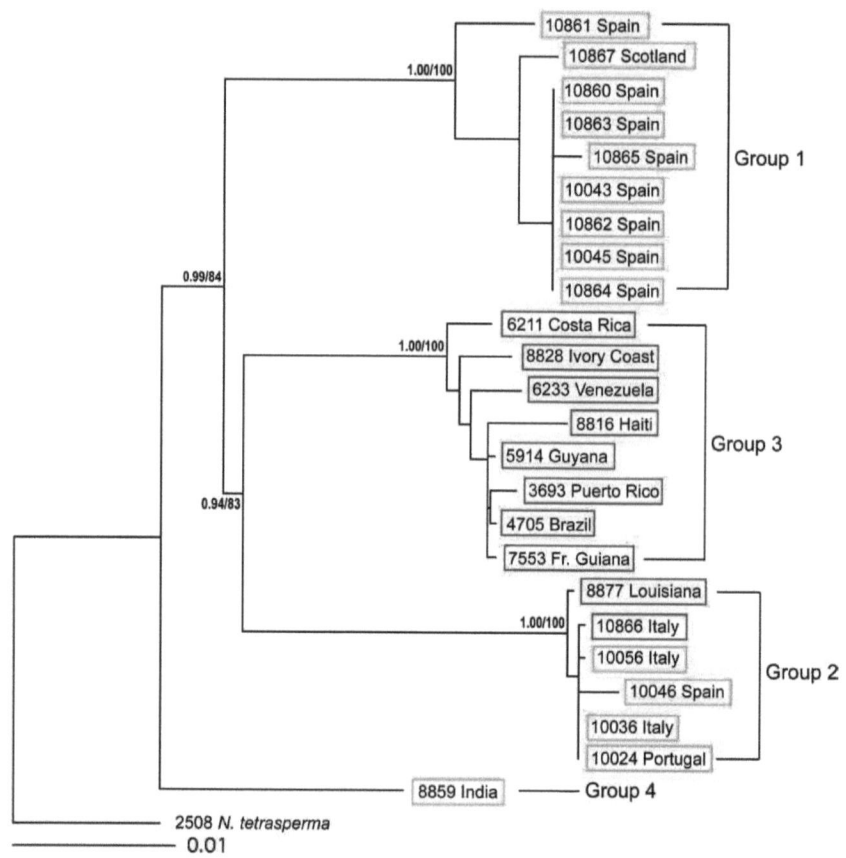

Figure 3.14: Maximum likelihood phylogram of *white collar-1* and *frequency* promoter combined for 24 *Neurospora crassa* isolates. Different colours depict different clades (red = clade A, blue = clade B, green = clade C). *N. tetrasperma* was taken as outgroup. Numbers near branches indicate confidence levels (Bayesian posterior probability/maximum likelihood bootstrap proportion); numbers in front of collection sites indicate the FGSC number. Bar indicates substitutions per site.

Spanish isolates (10860, 10862-64, 10043, 10045) did not show any genetic difference in clock genes and were therefore considered as one individual. Overall, there was a positive correlation between latitude of the origin and substitutions per site in *wc-1* gene and *frq* promoter combined analysis, meaning the more northern the isolates, more substitutions per site they have ($p = 0.0004$, $r = 0.7336$; Figure 3.15). Separating isolates into "test group" ($p = 0.1345$, $r = 0.5072$) or phylogenetic groups obtained with combined analysis there was no evident correlation (group 1: $p = 0.9860$, $r = 0.0140$; group 2: $p =$ ns, $r_S = 0.3582$; group 3: $p = 0.1449$, $r = 0.5645$; Figure 3.15). However, only isolates from clade A showed positive correlation ($p = 0.0055$, $r = 0.7993$; data not shown).

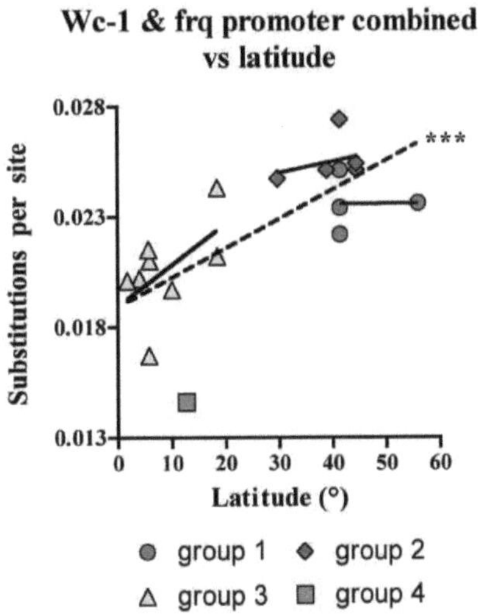

Wc-1 & frq promoter combined vs latitude

Figure 3.15: Correlation between latitude and substitutions per site in combined analysis of *wc-1* gene and *frq* promoter. Different groups are derived from phylogenetic analysis on *wc-1* and *frq* promoter combined. Dashed linear regression line refers to all isolates.

Considering phylogenetic groups obtained from the combined analysis, substitutions per site for *wc-1* gene and *frq* promoter were calculated (Figure 3.16). *Frq* promoter had more substitutions per site in the group 2 and 3 than *wc-1* gene (on average 0.046 vs 0.011 for group 2 and 0.038 vs 0.00 for group 3), while for the groups 1 and 4 the number of substitutions was similar in *wc-1* and *frq* promoter (0.021 vs 0.021 for group 1 and 0.014 vs 0.017 for group 4).

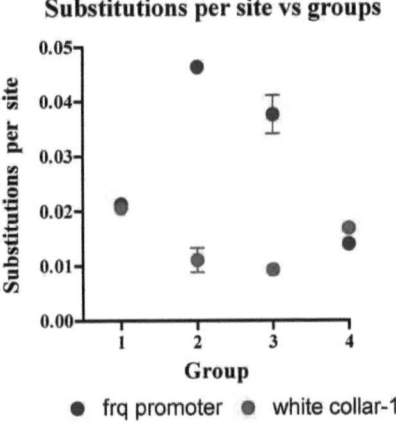

Figure 3.16: Substitutions per site for groups found in combined phylogenetic analysis of *wc-1* and *frq* promoter. *Frq* promoter is depicted in blue and *wc-1* gene in red. The dots represent average and bars standard deviations.

The average of substitutions per site for each sequenced part of *frq* promoter and *white collar-1* gene was calculated (Table 3.4). In *frq* promoter and *wc-1*, part I had more substitutions per site in each group than part II (0.024 vs. 0.017, 0.081 vs. 0.021, 0.018 vs. 0.11 for *frq* promoter, and 0.029 vs. 0.012, 0.016 vs. 0.006, 0.014 vs. 0.005, 0.019 vs. 0.015 for *wc-1*). Exception was group 3, which had more substitution per site in part II of the *frq* promoter (0.025 vs. 0.041).

Table 3.4: Mean number of substitutions per site ± standard deviation in DNA sequences of the *frq* promoter and *white collar-1* gene.

DNA sequence	Group (*wc-1* and promoter combined)			
	1 (N = 9 isol.)	2 (N = 6 isol.)	3 (N = 8 isol.)	4 (N = 1 isol.)
white collar-1 (part I & II)	.021 ± .000	.011 ± .002	.009 ± .001	.017 ± .000
part I	.029 ± .000	.016 ± .001	.014 ± .001	.019 ± .000
part II	.012 ± .001	.006 ± .004	.005 ± .002	.015 ± .000
frq promoter (part I & II)	.021 ± .001	.046 ± .000	.038 ± .003	.014 ± .000
part I	.024 ± .003	.081 ± .000	.025 ± .003	.018 ± .000
part II	.017 ± .002	.021 ± .000	.041 ± .003	.011 ± .000
All (promoter & *wc-1*)	.024 ± .001	.026 ± .001	.021 ± .002	.015 ± .000

The average number of repeats found in each group obtained with combined analysis of *frq* promoter and *wc-1* gene was calculated (Table 3.5). Group 2 had the longest 5'AG/GA and CpolyQH repeat from all groups (17.3 nucleotides and 35.8 amino acids), while group 4 had the longest NpolyQ (42 amino acids). The group 3 was the most variable one, with the mean standard deviation of 1.5.

Table 3.5: Mean number of SSRs ± standard deviation of the *white collar-1* gene.

Number of repeats	Group (*wc-1* and promoter combined)			
	1 (N = 9 isol.)	2 (N = 6 isol.)	3 (N = 8 isol.)	4 (N = 1 isol.)
5'AG/GA	13.0 ± 0	17.3 ± 1.63	14.0 ± 0.53	9.0 ± 0
NpolyQ	26.0 ± 0	35.8 ± 0.41	36.5 ± 3.21	42.0 ± 0
CpolyQH	5.0 ± 0	9.2 ± 0.41	8.5 ± 0.76	6.0 ± 0

Physiology

For the physiological analysis, 24 different *Neurospora crassa* isolates collected from all over the world were analysed in race tubes. Since *N. crassa* asexual spores (conidia) are produced in a circadian manner and become visible as dense bands when grown in race tubes, this can be used as output of the circadian clock. Therefore, the period and the phase of the conidiation rhythm in different light-dark (LD) or temperature conditions can be measured. As a control, standard laboratory strain bdA was used. This strain has period in constant darkness of 22 h and shows no banding in complete light. Here, a total of ca. 2500 and 1600 race tubes, for light and temperature conditions, respectively, were analysed. For each isolate, the free-running period (FRP) in complete darkness, period in entrained conditions (tau_E) and onset of conidiation (Phi on) relative to mid-dark/cold was calculated. Race tubes where the growth was arrhythmic, that is, bands of conidiation were not visible were excluded from the analysis.

The clock of isolates in constant conditions

In complete light all isolates showed arrhythmic growth, therefore, the free-running period was calculated in complete darkness (DD) at 25°C. From all isolates tested, 46% of them showed arrhythmic growth in DD, and were excluded from the analysis. From 13 isolates was possible to calculate the FRP as they showed clear visible banding. Cluster analysis performed on median FRP for each isolate gave three different clusters: cluster a corresponded to all isolates with arrhythmic growth, cluster 2 to isolates 10036 Italy, 10866 Italy, 10046 Spain and 10024 Portugal, and all the rest isolates formed cluster 1. Summarizing clusters according to clock-genes phylogenetic groups (Table 3.6), they did not separate nor to clade nor to clock-genes-phylogenetic groups. Thus, the FRP is not given by clade nor clock-genes genetic properties.

Table 3.6: FRPs and their clusters for each isolate. The isolates are arranged according to *wc-1* and *frq* promoter combined phylogenetic groups. The numbers in front of the collection sites indicate FGSC numbers.

Isolate	FRP cluster
10860 Spain	a
10862 Spain	a
10863 Spain	a
10864 Spain	F1
10043 Spain	a
10045 Spain	a
10865 Spain	a
10867 Scotland	F1
10861 Spain	a
10036 Italy	F2
10866 Italy	F2
10056 Italy	F1
10046 Spain	F2
10024 Portugal	F2
8877 Louisiana	F1
3693 Puerto Rico	F1
4705 Brazil	a
5914 Guyana	a
7553 Fr. Guiana	a
8816 Haiti	F1
6233 Venezuela	F1
8828 Ivory Coast	a
6211 Costa Rica	F1
8859 India	F1
1858 bdA	F1

a=arrhythmic

In complete darkness (DD), all individuals of the control strain *bdA* and of the cluster 2 (F2) formed rhythmic bands whereas cluster a isolates showed arrhythmic growth (Figure 3.17). The cluster 1 (F1) was the most variable one, where % of rhythmic individuals varied between 7% (10056 Italy) and 100% (isolates from Louisiana, Costa Rica, and India).

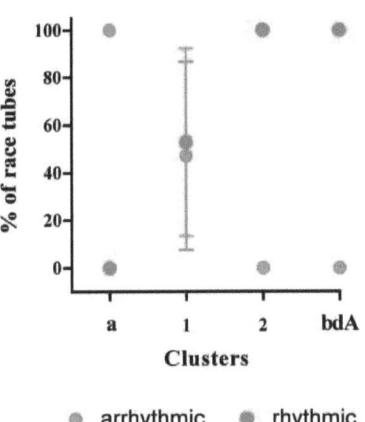

Figure 3.17: % of rhythmic and arrhythmic individuals in DD for each cluster. Dots represent average and bars standard deviation.

The range of FRPs varied between 22 – 32 h. Most of the isolates (62%) had their FRP around 23 h, only few of them had a FRP around 32 h (isolates 10036 Italy, 10866 Italy, 10046 Spain and 10024 Portugal) or around 22 h (10867 Scotland and 10864 Spain). Cluster 1 had a median free-running period of 23.1 h, while cluster 2 of 31.8 h. Furthermore, control strain *bdA* had an median FRP of 23.8 h (Figure 3.18). One-way ANOVA analysis with post hoc Dunn's multiple comparison test showed that the FRP of cluster 2 was significant different from the FRPs of all other groups ($p < 0.001$; Figure 3.18).

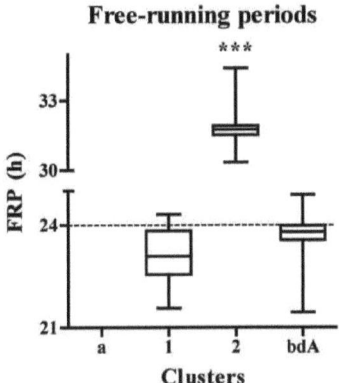

Figure 3.18: Distribution of FRPs according to the clusters. bdA was used as control. Median and range are given.

A positive correlation between FRPs and latitude of the isolates' origin was found; more northern the isolates, the longer the FRP (p < 0.0001, r_S = 0.4870; Figure 3.19). "Test group" showed no correlation (p = 0.3837, r_S = -0.1593), neither did phylogenetic groups obtained with *wc-1* and *frq* promoter combined analysis (group 1: p = 0.1793, r = -0.7097, group 2: p = 0.0595, r_S = 0.3314, group 3: p = 0.5284, r_S = -0.1644; data not shown). Also different FRP-clusters, showed no correlation (cluster 1: p = 0.0597, r_S = -0.3124; cluster 2: p = 0.7219, r_S = -0.0718; Figure 3.19).

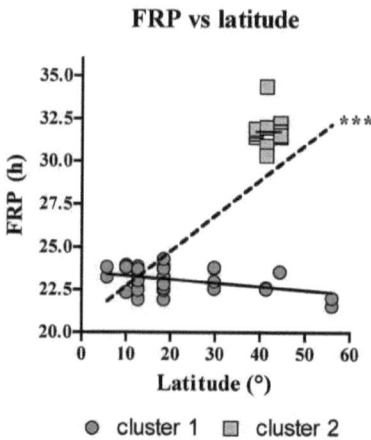

Figure 3.19: Correlation between FRP and latitude. Dashed regression line refers to all isolates.

The clock in entrained conditions

In order to investigate differences at physiological level between isolates coming from various locations, a series of circadian surface experiments have been conducted. For these experiments the cycle length was 24 hours, and the amount of light/warm and darkness/cold, for light/temperature surface, respectively, were changed. These experiments have short photo/thermo-periods (conditions with less than 50% of light/warm) and long photo/thermo-periods (conditions with more than 50% of light/warm). With the help of ChronOSX Program the period (which was used to test the entrainability) and phase of entrainment (Phi onset = beginning of the conidiation) relative to mid-dark/cold for each isolate was calculated. In the light-dark and temperature cycles, the isolates could be either entrained to the *zeitgeber's* cycles, arrhythmic, or rhythmic but not entrained. The last two possibilities were excluded from the analysis (see Materials and Methods).

Cluster analysis

The cluster analysis was performed with the median of the phase of entrainment for each isolate considering all light-dark cycles together and all temperature cycles together. In light surface, condition 20/4 LD was excluded from the analysis because only 39% of the isolates showed entrained conidiation. 10056 Italy and 5914 Guyana isolates were excluded from the analysis because they showed only entrained conidiation in 26% and 35% of all data analysed, respectively. For temperature surface, the conditions 75% and 84% of warm were excluded because only 54% and 32% of isolates showed entrained conidiation, respectively. Isolates 10860 Spain, 10043 Spain, and 5914 Guyana were excluded from the temperature surface analysis, with only 21%, 2%, and 27% of the race tubes that could be analysed, respectively.

The cluster analysis performed on median phase of entrainment for each isolate gave three clusters, both for light-dark and temperature cycles (Table 3.7). Most of the isolates have the same cluster belongings for light-dark and temperature cycles. Exception was 8859 India isolate, which in light belonged to cluster 3 and in temperature to cluster 1. The clusters found here did not respect nor clock nor clade phylogeny, thus the phase of entrainment could not be described by the clade or clock genetics.

Table 3.7: Cluster analysis of the phase of entrainment (PhiOn) for light and temperature surface. The isolates are arranged according to *wc-1* and *frq* promoter combined phylogenetic groups. The numbers in front of the collection sites are FGSC numbers

Isolate	Clusters	
	PhiOn light	PhiOn temp.
10860 Spain	L1	–
10862 Spain	L1	T1
10863 Spain	L1	T1
10864 Spain	L1	T1
10043 Spain	L1	–
10045 Spain	L1	T1
10865 Spain	L1	T1
10867 Scotland	L1	T1
10861 Spain	L1	T1
10036 Italy	L2	T2
10866 Italy	L2	T2
10056 Italy	–	–
10046 Spain	L2	T2
10024 Portugal	L2	T2
8877 Louisiana	L1	T1
3693 Puerto Rico	L3	T3
4705 Brazil	L1	T1
5914 Guyana	–	–
7553 Fr. Guiana	L1	T1
8816 Haiti	L3	T3
6233 Venezuela	L1	T1
8828 Ivory Coast	L1	T1
6211 Costa Rica	L1	T1
8859 India	L3	T1
1858 bdA	L1	T1

In the light-dark conditions, the phases of entrainment for cluster 1 isolates increased with increasing amounts of light during the short photoperiods (periods which have less than 50% of light) reaching the midnight and in long photoperiods remained around midnight. In the temperature cycles, phases of entrainment decreased more or less continuously with increasing amount of warm temperature from mid-cold to ca. -4 h (Figure 3.20).

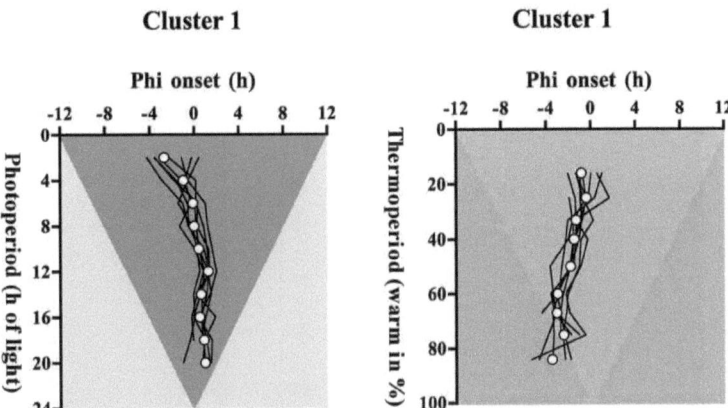

Figure 3.20: Light (left) and temperature (right) surface of cluster 1. The grey/blue area corresponds to dark/cold and yellow/red to light/warm portion of the cycle. X-axis shows phase of entrainment (Phi onset) in hours, while y-axis shows different photo/thermo-periods. The white dots represent median of all isolates in one condition, while the lines the median of each isolate.

In light-dark and temperature cycles, the phase of entrainment for cluster 2 isolates decreased with increasing amounts of light and temperature, respectively and occurred ca. 6 h after the midnight and mid-cold, respectively (Figure 3.21).

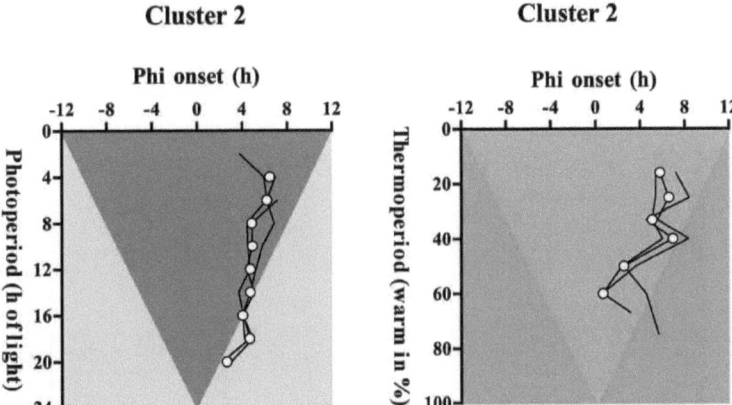

Figure 3.21: Light (left) and temperature (right) surface of cluster 2. The grey/blue area corresponds to dark/cold and yellow/red to light/warm portion of the cycle. X-axis shows phase of entrainment (Phi onset) in hours, while y-axis shows different photo/thermo-periods. The white dots represent median of all isolates in one condition, while the lines the median of each isolate.

The phase of entrainment of isolates belonging to cluster 3 increased with increasing amounts of light and ranged from midnight to ca. 3 h after midnight. In temperature cycles, phase of entrainment deceased with increasing warm portions and ranged from 1 h to ca. 4 h before mid-cold (Figure 3.22).

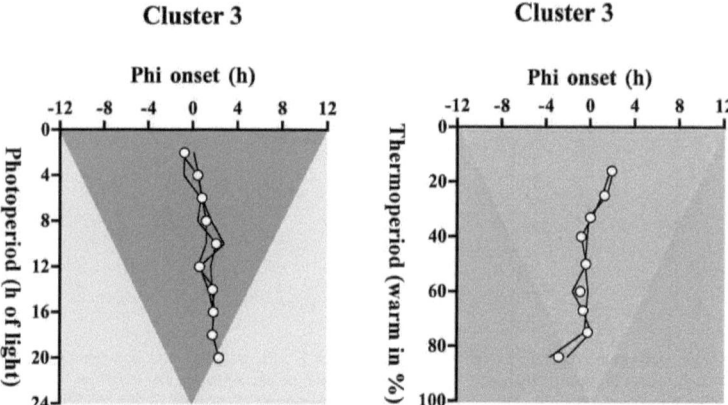

Figure 3.22: Light (left) and temperature (right) surface of cluster 3. The grey/blue area corresponds to dark/cold and yellow/red to light/warm portion of the cycle. X-axis shows phase of entrainment (Phi onset) in hours, while y-axis shows different photo/thermo-periods. The white dots represent median of all isolates in one condition, while the lines the median of each isolate.

The phase of entrainment for the control strain *bdA* increased with increasing amounts of light during the short photoperiods (from -4 h to midnight) and remained then rather constant. In temperature cycles the phase of entrainment became earlier with increasing warm amounts (Figure 3.23).

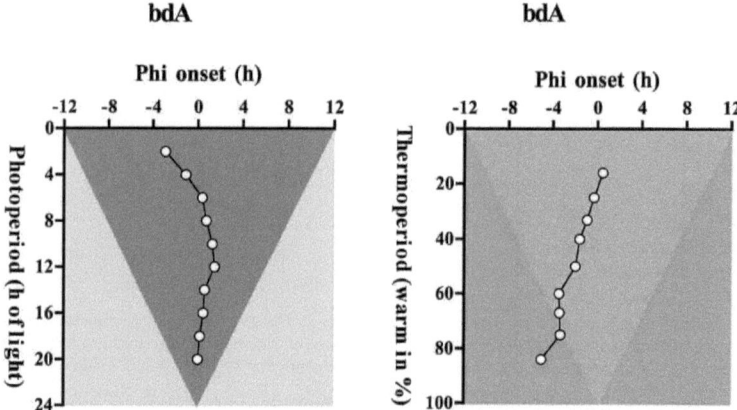

Figure 3.23: Light (left) and temperature (right) surface of *bdA*. The grey/blue area corresponds to dark/cold and yellow/red to light/warm portion of the cycle. X-axis shows phase of entrainment (Phi onset) in hours, while y-axis shows different photo/thermo-periods. The white dots represent median in each condition.

The phase of entrainment of the cluster 2 was significantly longer (median of 4.74 and 5.46 h in light and temperature surface, respectively) than in any of the other clusters (median of 0.47, 1.51 and 0.45 h for cluster 1, 3 and *bdA* in light surface, respectively; one-way ANOVAs with Bonferroni post-hoc tests: $p < 0.001$) both in light and temperature surface (Figure 3.24). Overall, in temperature cycles the phase of entrainment was earlier than in light-dark cycles for clusters 1, 3 and *bdA* (-1.77, -0.36 and -1.92 h, respectively).

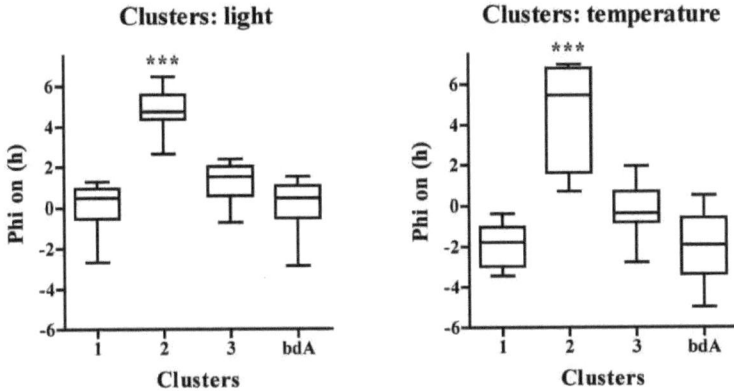

Figure 3.24: Median of phase of entrainment (Phi on) in light (left) and temperature (right) surface.

Overall, positive correlation between FRP and phase of entrainment in light ($p = 0.0007$, $r_S = 0.8132$) but not in temperature surface ($p = 0.0518$, $r_S = 0.5495$; Figure 3.25) was found. When separating isolates into clusters, the correlation was not evident any more (light: cluster 1: $p = 0.0975$, $r = 0.7329$; cluster 2: $p = 0.8742$, $r = -0.1258$; temperature: cluster 1: $p = 0.7858$, $r = -0.1272$; cluster 2: $p = $ ns, $r_S = 0.2000$; Figure 3.25). "Test group" showed no correlation between FRP and phase of entrainment (light: $p = 0.4595$, $r = -0.3784$; temperature: $p = 0.2607$, $r = -0.5476$; data not shown).

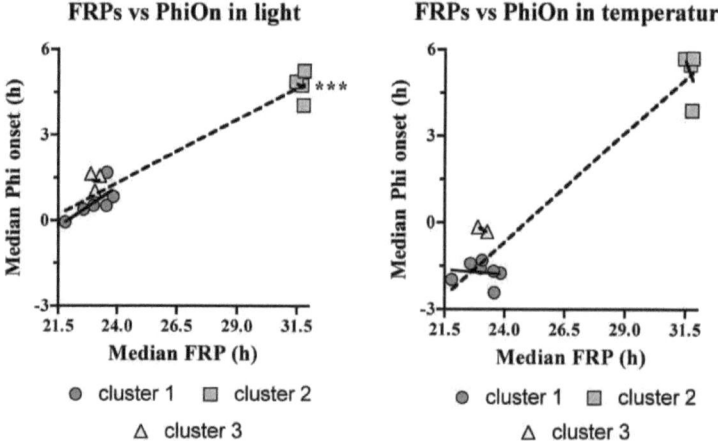

Figure 3.25: Correlation between FRPs and phase of entrainment (Phi onset) in light (left) and temperature (right) surface. Dashed linear regression line refers to all isolates.

In order to give not too much weight to Spanish isolates, from all of them, except 10046, the mean was calculated for the latitudinal cline analysis. There was no correlation between phase of entrainment and latitude of the origin in light (cluster 1: $p = 0.9323$, $r = -0.03098$; cluster 2: $p = 0.1810$, $r = -0.8190$; all: $p = 0.0884$, $r_S = 0.4258$) and temperature surface (cluster 1: $p = 0.4157$, $r = -0.2904$; cluster 2: $p = $ ns, $r_S = -0.8000$; all: $p = 0.3023$, $r_S = 0.2752$; Figure 3.26). "Test group", also, showed no correlation (light: $p = 0.3212$, $r = 0.3502$; temperature: $p = 0.7711$, $r = -0.0994$; data not shown).

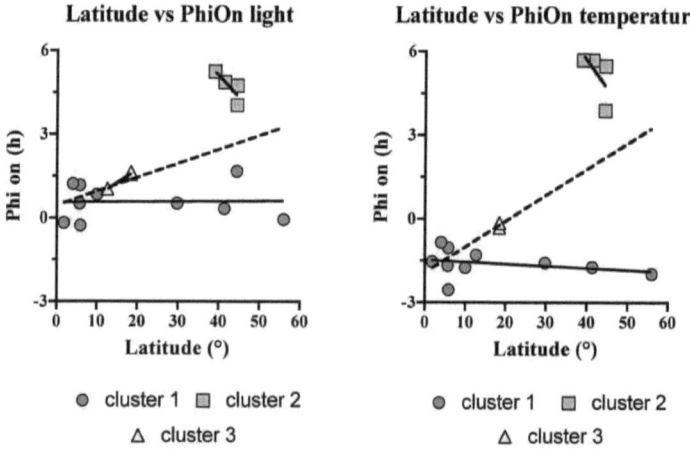

Figure 3.26: Correlation between latitude and phase of entrainment (Phi on) in light (left) and temperature (right) surface. Dashed linear regression line refers to all isolates.

Variability of the data

Both in light and temperature surface, isolates showed very variable data. In light surface none isolate showed conidiation bands in 22/2 LD condition, i.e., they were all arrhythmic. In 20/4 LD only 38% of the isolates were showed rhythmic entrained growth. In other L/D conditions, there were between 75% (4/20 LD) and 100% (8/16 LD and 10/14 LD) of isolates with rhythmic growth. From these, entrained were between 71% (4/20 LD) and 96% (6/18 LD, 8/16 LD, 10/14 LD, 12/12 LD and 16/8 LD) of the isolates. The most arrhythmic isolate was 10056 Italy with 72% of race tubes been without visible bands. The best-entrained isolate was from Venezuela with 86% of conidiation bands perfectly entrained to *zeitgeber* cycle. In temperature surface, there were between 54% (condition with 84% of warm) and 88% (50% of warm) isolates with rhythmic conidiation, from which only 33% entrained in condition with 84% of warm and 88% in condition with 50% of warm. The most arrhythmic isolate was again 10056 Italy, which did not show visible banding in any condition, and the best-entrained isolate was from Puerto Rico with 91% of entrained conidiation bands.

To test how isolates stably entrained in different light-dark and temperature cycles, the period (tau_E) was calculated. The distribution of tau_E was quite broad and was larger in temperature than in light-dark cycles (ranged from 20.36 – 35.85 h vs. 20.66 – 32.60 h). The median tau_E was 23.88 h for wild-type isolates and 23.96 h for *bdA* in light surface, while 23.89 h for wild-type isolates and 23.87 h for *bdA* in temperature surface (Figure 3.27). Furthermore, most of the isolates had periods between 23 - 24 h (82% of isolates in light and 77% of isolates in temperature surface).

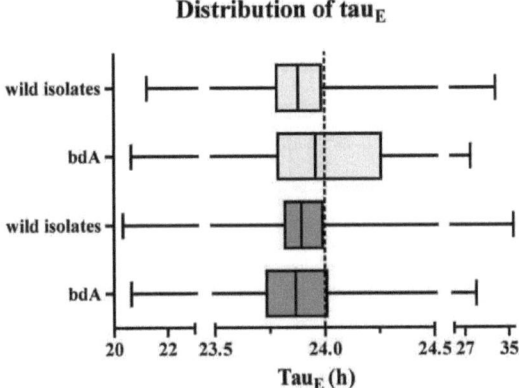

Figure 3.27: Distribution of periods in light-dark (yellow) and temperature cycles (reddish).

Also here as for the phase of entrainment, the mean of all Spanish isolates, except 10046, was calculated and they are represented as one individual. The percentage of race tubes with entrained conidiation for all isolates correlated with latitude in light (p < 0.0001, r = -0.8645) and temperature (p = 0.0003, r = -0.8089) cycles. If the isolates were divided into clusters, only cluster 1 showed same correlation (light: p = 0.0070, r = -0.8536; temperature: p = 0.0052, r = -0.8336; Figure 3.28). "Test group" showed correlation in light surface (p = 0.0354, r = -0.7009) but not in temperature (p = 0.2245, r = -0.4497; data not shown).

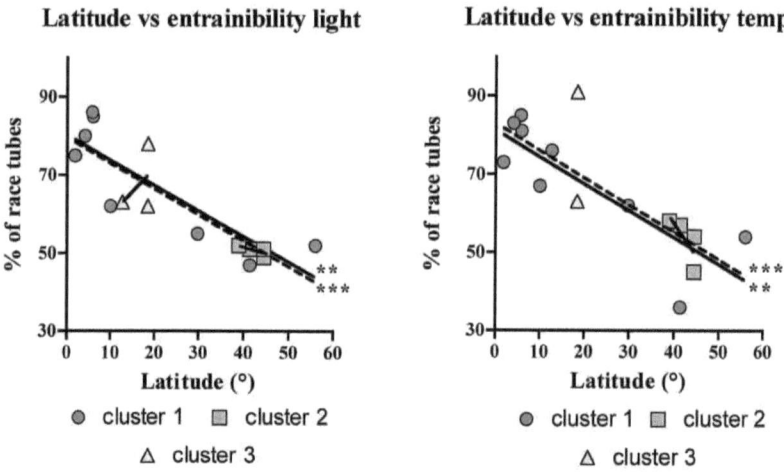

Figure 3.28: Correlation between race tubes showing entrained conidiation in light (left) and temperature (right) surface with latitude. Dashed regression line refers to all isolates.

Comparison between light and temperature surface

In order to test the correlation between data obtained in light and temperature surface, a linear regression was calculated. The overall number of race tubes with arrhythmic and entrained conidiation in light versus temperature surface showed positive correlation (arrhythmic: p < 0.0001, r_S = 0.7685; entrained: p < 0.0001, r = 0.7859; data not shown). Meaning that the more race tubes with arrhythmic/entrained conidiation in light surface, the more race tubes with arrhythmic/entrained conidiation, also, in temperature surface. Thus, the isolates behaved similar in cycles with light and temperature as *zeitgeber*. Considering different clusters, only cluster 1 showed the same correlations (arrhythmic: p = 0.0021, r = 0.6743; entrained: p = 0.0008, r = 0.7355; data not shown) as did the "test group" (arrhythmic. p = 0.0174, r_S = 0.7446, entrained: p = 0.0001, r = 0.9254; data not shown). Furthermore, the phase of entrainment in temperature surface correlated with phase in light surface, meaning that isolates showed similar behaviour in light and temperature surface (p < 0.0001, r_S = 0.8063; Figure 3.29). Only cluster 1 isolates (p = 0.0131, r = 0.6047; Figure 3.29) and "test group" (p = 0.0007, r = 0.8822; data not shown) showed positive correlation.

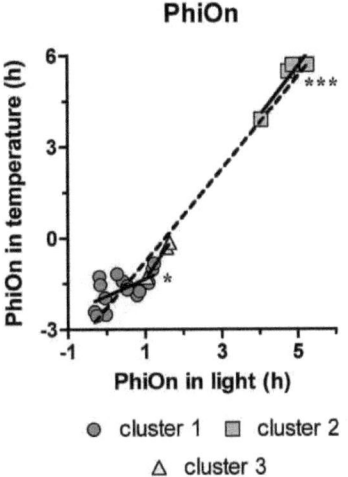

Figure 3.29: Correlation between phases of entrainment (Phi on) in light vs. temperature surface. Dashed linear regression line refers to all isolates.

Association between DNA sequences and circadian phenotypes

The aim of this part was to see if the physiology of the isolates could be explained by their genetic properties. Therefore, a correlation between phylogenetic and physiological results was calculated.

Free-running periods

Michael *et al.* (2007) found that the number of repeats in WC-1 NpolyQ domain correlated with period length, the same correlation was tested here. Here no overall correlation between FRP and number of repeats in WC-1 NpolyQ domain (p = 0.2555, r$_S$ = 0.1442; Figure 3.30) was found. "Test group" also showed no correlation between FRP and NpolyQ repeat (p = 0.5323, r$_S$ = 0.1127; data not shown), nor if the isolates were separated according to the FRP-clusters (cluster 1: p = 0.0718, r$_S$ = 0.2994; Figure 3.30). Separating isolates into phylogenetic groups found with *wc-1* and *frq* promoter combined analysis, only groups 2 showed correlation between length of NpolyQ and FRP (p = 0.0002, r$_S$ = 0.6036; data not shown).

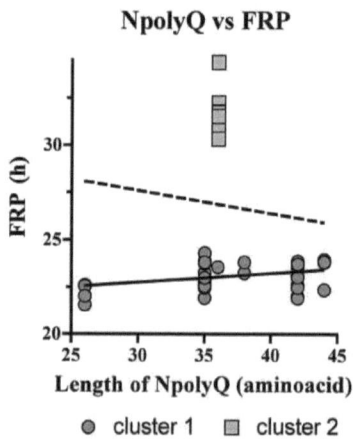

Figure 3.30: Correlation between number of repeats in WC-1 NpolyQ and FRP. Dashed regression line refers to all isolates.

The number of repeats in WC-1 NpolyQ domain was grouped in three lengths (26, 35 and >36 aminoacids) for isolates with arrhythmic and rhythmic conidiation. Mann Whitney test showed that rhythmic isolates had significantly longer repeats in WC-1 NpolyQ domain than the arrhythmic ones (p < 0.0001; Figure 3.31).

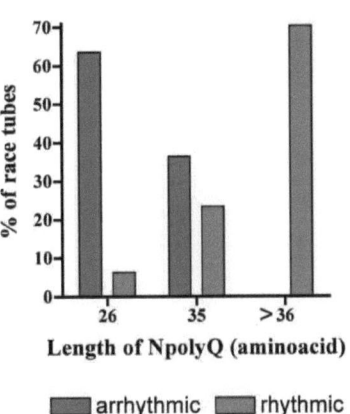

Figure 3.31: Number of repeats in WC-1 NpolyQ and their distribution in rhythmic/arrhythmic isolates.

Next, the correlation between FRP and substitutions per site in clock genes was calculated. Overall, FRP showed a correlation with substitutions per site of *frq* promoter (p < 0.0001, r_S = 0.6971; data not shown) and *wc-1-frq* promoter combined (p < 0.0001, r_S = 0.7221; Figure 3.32) but not with substitutions per site of *wc-1* gene (p = 0.6890, r_S = -0.0510; data not shown). When isolates were divided into FRP-clusters, FRP of cluster 1 isolates showed a correlation with substitutions per site of *wc-1* gene (p = 0.0369, r_S = -0.3443) but not with substitutions per site of *frq* promoter (p = 0.3710, r_S = 0.1514) and *wc-1-frq* promoter combined (p = 0.9434, r_S = 0.0121). "Test group" showed no correlation between FRP and substitutions per site in any clock gene (*frq* promoter: p = 0.8703, r_S = 0.0301; *wc-1* gene: p = 0.4078, r_S = -0.1515; *wc-1-frq* promoter combined: p = 0.8606, r_S = 0.0323; data not shown).

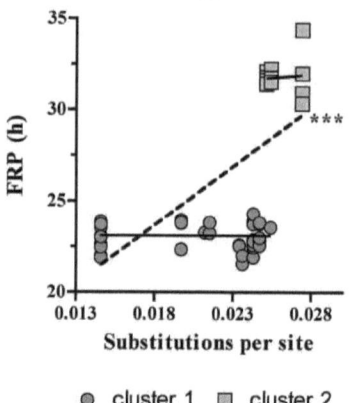

Figure 3.32: Correlation between FRP and substitutions per site in *wc-1-frq* promoter combined. Dashed linear regression line refers to all isolates.

Clock in entrained conditions

To see if the properties of the isolates, such as entrainibility and arrhythmicity, were influenced by their genetics, the linear correlation between percentage of the race tubes with entrained/arrhythmic growth and the number of repeats in WC-1 NpolyQ domain was calculated. Here, the mean of all Spanish isolates (except 10046) was calculated. No correlation between number of race tubes with entrained growth and length of NpolyQ in light and temperature surface was found (light: $p = 0.3398$, $r = 0.2554$; temperature: $p = 0.1312$, $r = 0.3939$; Figure 3.33). The same was true for the race tubes with arrhythmic growth (light: $p = 0.3395$, $r = -0.2469$; temperature: $p = 0.1141$, $r_S = -0.3975$; data not shown). If the isolates were separated into "test group" no correlation was found in race tubes with entrained (light: $p = 0.8336$, $r = -0.0765$; temperature: $p = 0.8238$, $r = 0.0811$) nor with arrhythmic growth (light: $p = 0.6123$, $r = 0.1833$; temperature: $p = 0.5603$, $r_S = -0.2038$) in both light and temperature surface. When isolates were separated according to the PhiOn-clusters, no correlation between length of NpolyQ domain and race tubes with entrained growth was found (light: $p = 0.3114$, $r = 0.3812$; temperature: $p = 0.1441$, $r = 0.4967$; Figure 3.33). Considering race tubes with arrhythmic growth of cluster 1 no correlation with length of NpolyQ domain was found (data not shown; light: $p = 0.2153$, $r = -0.4578$; temperature: $p = 0.1210$, $r = -0.5549$).

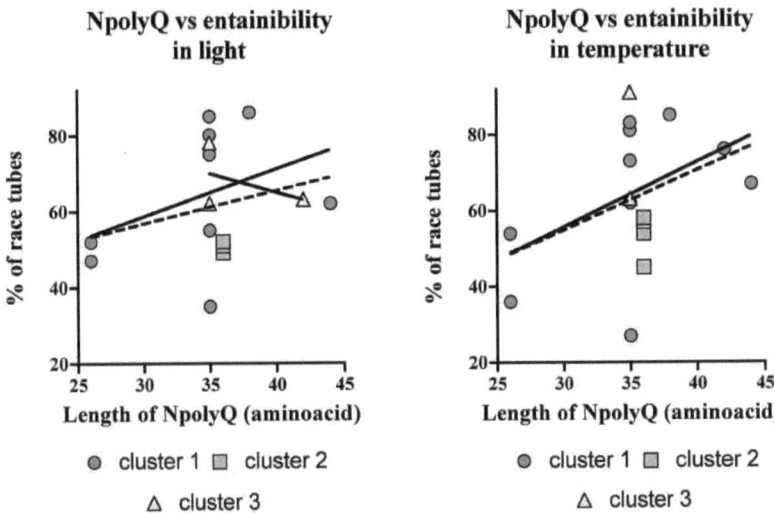

Figure 3.33: Correlation between length of WC-1 NpolyQ with race tubes with entrained conidiation in light (left) and temperature (right) surface. Dashed linear regression line refers to all isolates.

For correlation between entrainibility (% of the race tubes with entrained growth) and substitutions per site in clock genes, the mean of Spanish isolates (10860, 10862-64, 10043, 10045) was calculated. Considering all isolates together (data not shown), entrainibility correlated with substitutions per site in *wc-1* gene in temperature (p = 0.0365, r_S = -0.4824) but not in light surface (p = 0.2107, r = -0.3008). No correlation between entrainibility and substitutions per site in *frq* promoter (light: p = 0.8447, r_S = -0.0482; temperature: p = 0.8076, r = -0.0599) and *wc-1-frq* promoter combined was found (light: p = 0.1251, r = -0.3643; temperature: p = 0.1422, r = -0.3497; data not shown). Entrained isolates from light surface cluster 1 did correlate with substitutions per site in *wc-1* gene (p = 0.0347, r_S = -0.6770) and in *wc-1-frq* promoter combined (p = 0.0095, r = -0.7681) but not in *frq* promoter (p = 0.0805, r = 0.5774; data not shown). "Test group" did not correlate in any gene (*frq* promoter: p = 0.9817, r = 0.0084; *wc-1* gene: p = 0.9528, r = -0.0216; *wc-1-frq* promoter combined: p = 0.4834, r = 0.2515; data not shown). Entrained isolates from temperature surface cluster 1 did correlate with substitutions per site in *wc-1* gene (p = 0.0204, r_S = -0.6943) and in *wc-1-frq* promoter combined (p = 0.0487, r = -0.6049) but not in *frq* promoter (p = 0.1339, r = 0.4813; data not shown). "Test group" in temperature surface did not correlate in any gene (*frq* promoter: p = 0.9730, r_S = -0.0182; *wc-1* gene: p = 0.7330, r_S = -0.1297; *wc-1-frq* promoter combined: p = 0.5743, r = 0.2027; data not shown).

An overall correlation between arrhythmicity (% of race tubes with arrhythmic growth) in light surface with substitutions per site in *wc-1* gene (p = 0.0073, r_S = 0.5942) but not in substitutions per site of *frq* promoter (p = 0.2181, r = -0.2963) and *wc-1-frq* promoter combined (p = 0.7602, r = 0.0750) was found. If the isolates were separated into "test group" correlation was not evident (*frq* promoter: p = 0.1912, r_s = -0.4512; *wc-1* gene: p = 0.6321, r_s = 0.1734; *wc-1-frq* promoter combined: p = 0.0681, r = -0.5975), however cluster 1 showed correlation with substitutions per site in all genes (*frq* promoter: p = 0.0066, r = -0.7898; *wc-1* gene: p = 0.0149, r_s = 0.7509; *wc-1-frq* promoter combined: p = 0.0366, r = 0.6632). Overall, arrhythmicity in temperature surface showed no correlation with substitutions per site in any gene (*frq* promoter: p = 0.3111, r = -0.2455; *wc-1* gene: p = 0.1811, r_s = 0.3204; *wc-1-frq* promoter combined: p = 0.7894, r = 0.0657). "Test group" showed no correlation (*frq* promoter: p = 0.9730, r_s = 0.0123; *wc-1* gene: p = 0.9460, r_s = 0.0313; *wc-1-frq* promoter combined: p = 0.6649, r = -0.1570) but PhiOn cluster 1 did with substitutions per site in *wc-1* gene (p = 0.0182, r_s = 0.7053) and in *wc-1-frq* promoter combined (p = 0.0423, r = 0.6189, data not shown) but not in *frq* promoter (p = 0.0857, r = -0.5409; data not shown).

Next, the correlation between phase of entrainment and number of repeats in WC-1 NpolyQ domain was calculated. Here, the mean of all Spanish isolates (except 10046) was calculated and used in the analysis. No overall correlation between phase and the number of repeats in WC-1 NpolyQ domain in light (p = 0.1236, r_s = 0.4012) and temperature (p = 0.2739, r_s = 0.2912) surface was found (Figure 3.34). If the isolates were separated into the PhiOn-clusters, the correlation was, also, not significant (light: cluster 1: p = 0.3190, r = 0.3757; temperature: cluster 1: p = 0.5709, r = 0.2045; Figure 3.34) nor in isolates from the "test group" (light: p = 0.8912, r = 0.0498: temperature: p = 0.4688, r = -0.2596; data not shown).

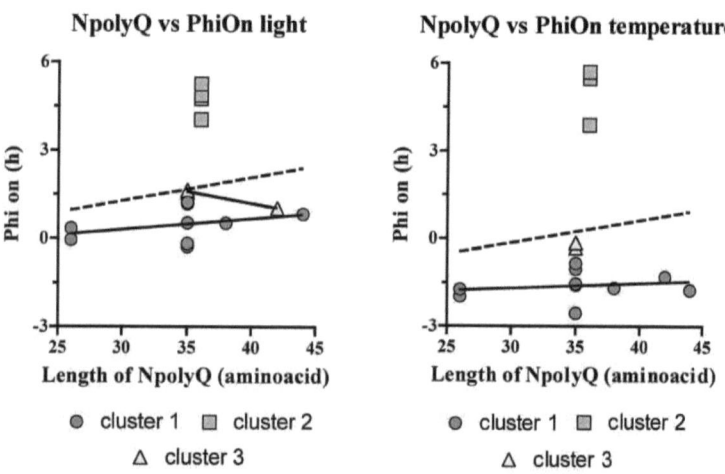

Figure 3.34: Correlation between length of WC-1 NpolyQ and phase of entrainment (Phi on) in light (left) and temperature (right) surface. Dashed linear regression line refers to all isolates.

Furthermore, overall correlation between phase of entrainment in light and temperature surface with substitutions per site in *frq* promoter (light: p = 0.0010, r_S = 0.6926; temperature: p = 0.0042, r_S = 0.6406; data not shown) and in *wc-1-frq* promoter combined (light: p = 0.0364, r_S = 0.4826; temperature: p = 0.0394, r_S = 0.4891; Figure 3.35) but not in *wc-1* gene (light: p = 0.9317, r_S = 0.0217; temperature: p = 0.9551, r_S = -0.0148; data not shown) was found. No correlation between phase and substitutions per site in any gene considering PhiOn-clusters was found (*wc-1* gene: light: p = 0.6193, r = -0.2090; temperature: p = 0.8140, r = 0.0856; *frq* promoter: light: p = 0.4344, r = 0.2990; temperature: p = 0.9480, r = -0.0223; *wc-1-frq* promoter combined: light: p = 0.5739, r = 0.1909; temperature: p = 0.6272, r = -0.1653; Figure 3.35). "Test group" showed, only in temperature surface, correlation between phase and substitutions per site in *wc-1-frq* promoter combined analysis (p = 0.0414, r_S = 0.5713) but not in single genes (*wc-1* gene: light: p = 0.4684, r_S= 0.2443; temperature: p = 0.8179, r_S = 0.07835; *frq* promoter: light: p = 0.3713, r_S = 0.3014; temperature: p = 0.2993, r_S = 0.3425; *wc-1-frq* promoter combined: light: p = 0.3242, r_S = 0.2971; data not shown). Here, the mean of Spanish isolates 10860, 10862-64, 10043, and 10045 for *frq* promoter and *wc-1-frq* promoter combined analysis and of all Spanish isolates, except 10046, for *wc-1* gene was calculated and used in correlation analysis.

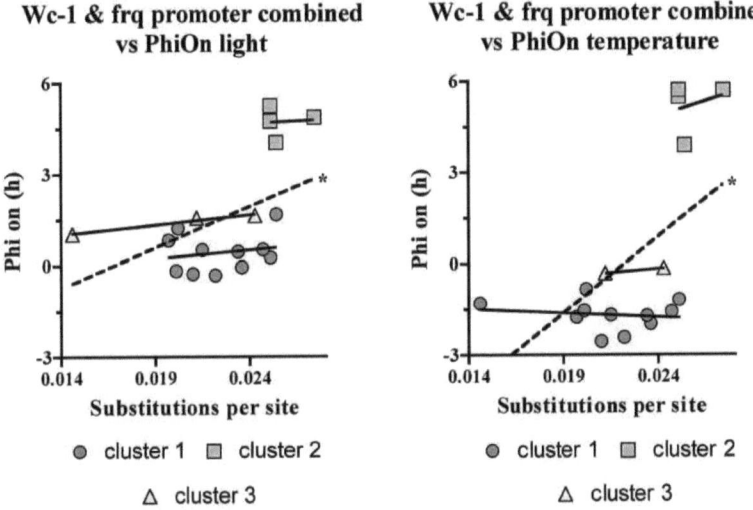

Figure 3.35: Correlation between phase of entrainment (Phi on) and substitutions per site in *wc-1-frq* promoter combined. Dashed linear regression line refers to all isolates.

4. DISCUSSION

Biological rhythms, as a result of the interplay between endogenous clocks and the environment, are very important properties of almost all organisms. The environment changes according to the time of day and year, creating therefore temporal information to the organisms. Such temporal information is registered by the clock, which "tells" the organism when to make use of the favourable season, when to avoid the effects of the unfavourable season and when to switch in a timely manner between the two lifestyles. The changing of season is predictable at any given latitude; with increasing latitude the favourable season becomes shorter, winter arrives earlier and spring arrives later. Hence, it is not surprising that a wide variety of organisms of all phyla use day length to anticipate the changing seasons to adjust their behaviour, development and reproduction. Understanding how the clock behaves at different latitudes provides a mean to understand the mechanism of geographical adaptation.

Neurospora crassa is a fungus found all over the world (Figure 1.16), whose circadian clock is well characterized and shows similarity to those of animals. Therefore, it appears to be a well-suited model to study mechanisms of environmental adaptation. The aim of this thesis was to see whether differences in circadian phenotype and genetics of wild-collected isolates of *Neurospora crassa* correspond to their geographical location and, more specially, whether they show a latitudinal cline.

Grouping of the isolates

Genetic and physiological analyses of 24 isolates showed that isolates belong to different groups according to their similarity in genetic sequences and circadian phenotype, respectively. There were three different clades found by sequencing of neutral markers, four and three phylogenetic groups (clock-gene groups) by sequencing the *white collar-1* gene and the *frequency* promoter, respectively, and two and three physiological clusters according to their free-running period and phase of entrainment in different light-dark and temperature cycles, respectively.

Geographical distribution of isolates across clades

The phylogenetic analysis of the four neutral markers showed the 24 isolates from different geographical locations to belong to the following clades: isolates from Caribbean Basin to clade A, most European isolates to clade B, one isolate (8859 India) to clade C. This distribution was in accordance with the analysis from Dettman *et al.* (2003) and Jacobson *et al.* (2006). In addition, two isolates (from Pakistan and Thailand) were found to form a new group basal to all clades, and one European isolate (10866 Italy) was unexpected found to belong to clade A rather than to clade B.

The latter findings suggest that there is no strict geographical distribution of clades. In fact, Dettman *et al.* (2003) found isolates from Ivory Coast to belong both to clade A and B, and

Jacobson et al. (2006) found the geographical distribution of clade B spanning from Europe to western North America, although fewer clade B isolates were found in North America.

It has been hypothesized that an Indian population gave rise to modern *N. crassa*. Subsequent migrations of *N. crassa* from India to Africa and the Caribbean Basin permitted differentiation of the Afro-Caribbean and Indian clades. Clade A is widespread across the Caribbean Basin and Africa. Clade C is limited to Tamil Nadu in India, whereas clade B comprises isolates from equatorial Africa, southern India, Europe and western North America (Turner et al., 2010). Still, clades A and B appear to coexist in equatorial Africa and Italy, which means that the *Neurospora* clades have no restricted geographical distribution.

However, there is no ready explanation for the appearance and evolution of clade B, which has members from Europe, North America, Africa and India (Jacobson et al., 2006) and appeared according to the phylogenetic tree as the latest clade. One possibility is that humans have influenced the geographical distribution of this clade by transporting it to different continents and releasing it into the environment. This holds also for the finding that the 10866 Italy isolate is an exception and indeed belongs to a different clade than 10056 Italy, although both isolates were collected from the same location.

Phylogenetic analysis of clock genes: Comparison with clades

To determine the genetic variability of the 24 isolates, two parts of the clock gene *white collar-1* and of the *frequency* promoter were sequenced. The phylogenetic analysis of the *wc-1*, of the *frq* promoter or of both clock components combined (*wc-1* + *frq* promoter) showed no exact correspondence with clade groups, as the phylogenetic trees obtained from sequences of neutral markers and clock components had different topologies (Figure 3.7 and 3.11). In fact, none of the three clades to which the isolates belong was exclusively represented in one of the clock-genes groups. Most of the isolates from clade B formed one group and most of the clade A isolates formed another distinct group in clock-genes tree. However, some clade-A isolates (10866 Italy and 8877 Louisiana, and 6233 Venezuela in *wc-1* tree) are found together with clade-B isolates (10024 Portugal, 10046 Spain, and 10036 and 10056 Italy) in one phylogenetic group of the clock genes. Interestingly in the *frq* promoter, the only clade-C isolate (India) is sister-group to this "mixed" group. On the other hand, in the *wc-1* tree the isolate from India formed one distinct group, which was basal to all other groups.

These results present the first systematic findings and phylogenetic analysis of two clock genes in *Neurospora crassa* isolates. Although the sample size was rather small, the results are supported by the fact that *wc-1* and *frq* promoter trees showed rather similar topologies (Figure 3.7 and 3.11).

Recently, Ellison *et al.* (2011) discovered two cryptic populations, one in the tropical Caribbean Basin and other in subtropical Louisiana, by sequencing the whole transcriptome of isolates from Caribbean Basin, South America and Africa, and concluded that the divergence island found between these two populations were due to the local adaptation. Results presented here support their findings, as the Louisiana and Caribbean Basin isolates did form two different groups. Since the European isolates formed another group, one might think that there could be three cryptic populations, one corresponding to the European isolates.

Given the different geographical distribution of clock-genes groups and clades, this may mean that clock genes and neutral markers have different evolutionary pressures. This idea is supported by their completely different functions. *Wc-1* is photoreceptor and a transcription factor, which binds to the *frq* promoter and starts the circadian feedback loop of *Neurospora*, while the neutral markers do not have a function or code for genes involved in metabolism (TML gene, Kaufman and Broquist, 1977). Thus, the differentiation of the clock components and clades in *Neurospora crassa* may have occurred *via* different evolutionary mechanisms.

Physiology

The free-running period (FRP) in complete darkness (DD) and the phase of entrainment to nine different light-dark and seven temperature cycles were used to group the isolates according to their similarity in circadian phenotype.

The analysis of FRP gave three different clusters: one corresponding to arrhythmic isolates (which did not show bands), one which corresponded to isolates whose FRP was around 23 h (cluster 1) and one with a FRP ranging around 32 h (cluster 2). Considering the phase of entrainment for both light and temperature surface, there were also three distinct clusters: cluster 1 corresponded to most of the isolates, cluster 2 comprised isolates from Italy, 10046 Spain and 10024 Portugal, and cluster 3 isolates from Puerto Rico, Haiti, and from India, which entrained only to light.

Since the phase of entrainment of cluster 1 and 3 did not differ significantly and cluster 2 appeared to be significantly different from the rest (Figure 3.24), there were overall only two different phenotypic groups. Group 1 comprised isolates with short free-running period (FRP < 24h) and group 2 with a long FRP (> 24h). Both groups also differed in the phase of entrainment: group 1 had an onset of conidiation locked to midnight, while group 2 had an onset of conidiation shifted to later hours. Remi *et al.* (2010) found a similar result by studying long- and short-period mutants.

An interesting case was the 10046 Spain isolate, which belonged to group 2 and not to group 1 like the other eight Spanish isolates. This may have resulted from a lack of genetic transfer from other members of population allowing mutation to accumulate slower or faster than in the other Spanish isolates. Furthermore, a preliminary study by Jacobson *et al.* (2006) assessing clonality in

Neurospora isolates from the same plant revealed more than one genetically distinct individual per plant. This supports the 10046 Spain result as more than one genetic individual can be present in very close spatial scale (Powell *et al.*, 2003).

Comparison between physiology and genetics

Since only two distinct physiological phenotypes were identified, a match with particular clades or particular clock-genes groups was rather unlikely. However, all members of the physiological cluster 2 (long-FRP phenotypes) belonged to clade B and to clock-genes group 2, but the reverse was not true, that is, not all members of clade B and clock-genes group 2 showed the long-FRP phenotype. Scotland and Spanish isolates did form one clock-genes group and one physiological cluster. Clock-genes group 2 consisted of isolates from Italy, 10046 Spain, 10024 Portugal and 8877 Louisiana, but the physiological cluster 2 only of Italian isolates, 10046 Spain and 10024 Portugal. Isolates from clock-genes group 3 were dispersed between physiological cluster 1 and 3.

Since, phylogenetic and physiological analysis do not coincidence, the differences found in circadian phenotype cannot be explained by genetic sequences of *wc-1* and *frq* promoter. Thus, sequencing of only two clock genes might not be sufficient for the analysis of circadian phenotypes. For example, as it was recently demonstrated, even when the same gene influences phenotypic variation in a given trait, the molecular mechanisms underlying this variation might differ between closely related species (Rosenblum *et al.*, 2010). Thus, different regulation mechanisms may be used to adjust life strategies in different environments.

Latitudinal cline

Latitudinal cline is a gradual change in a character over the geographical range of a species or population and is often associated with changes in environmental variables, such as the change of day length with latitude (Kyriacou *et al.*, 2007). In search for latitudinal clines in *Neurospora* isolates, several correlation analyses were performed: (1) regarding the number of substitutions per site in four neutral marker loci, (2) regarding the number of substitutions per site in the *wc-1* gene and the *frq* promoter locus, and regarding the number of repeats in the *wc-1* gene, and finally (3) in the free-running period (FRP) and the phase of entrainment. Furthermore, a possible existence of latitudinal clines was investigated (4) among the three different clades, (5) among the four phylogenetic groups of the *wc-1* gene and three phylogenetic groups of the *frq* promoter locus, and (6) among three clusters or two phenotype groups derived from the FRP and phase of entrainment.

Genetics

There was a positive correlation between substitutions per site and latitude in two of four neutral marker loci (TMI and DMG). This correlation was mainly due to the northern distribution of the

European isolates (clade B) reflecting their later phylogenetic origin than that from other clades. However, no latitudinal cline was found within the clade B or any other clade, with the exception of the clade A of the TMI and TML loci, which showed a positive correlation between substitutions per site and latitude – but this finding may be fortuitous as clade A included the one isolate (10866 Italy).

A positive correlation of substitutions per site and latitude was also found in *wc-1* gene, again reflecting a later evolution of clade B. The fact that the *frq* promoter does not show this correlation suggests that different selection pressures acted in these two genetic regions. Whereby the *wc-1* gene must have been under more specific selection based in the higher variability of the *frq* promoter. Furthermore, no correlation with latitude was evident in these two regions when isolates were separated into phylogenetic groups; thus isolates with same genetic background do not show a latitudinal cline.

Furthermore, the NpolyQ domain shows latitudinal clines in other organisms. The most thoroughly studied latitudinal cline is found in threonine–glycine (Thr-Gly) repeats of the clock gene *period* (*per*) of *Drosophila melanogaster* (Sawyer *et al.*, 1997). There, in northern Europe, the frequency of *per* (Thr-Gly)$_{20}$ increases, whereas that of *per* (Thr-Gly)$_{17}$ decreases with latitude. *Per* (Thr-Gly)$_{20}$ allele is probably better adapted in the colder and more thermally variable higher latitudes, whereas *per* (Thr-Gly)$_{17}$ is suited to the warmer Mediterranean region (Zamorzaeva *et al.*, 2005). Another clock protein of higher eukaryotes, CLOCK, has a polyglutamine repeat motif (PolyQ) at the carboxyl-terminal, which regulates its transcription-activating potential (Darlington *et al.*, 1998). Geographical surveys of *clock* gene variation in European blue tit and Chinook salmon have found evidence for a latitudinal cline in the PolyQ repeat length, with longer repeats found in higher latitudes (Johnsen *et al.*, 2007; O'Malley and Banks, 2008), and experimental studies in mice showed that expansion or contraction in the number of glutamine repeats in this region directly affects the corresponding gene product and influences the circadian phenotype (Vitaterna *et al.*, 1994).

The findings presented here do not support the results by Michael *et al.* (2007), who showed that the sizes of all three WC-1 SSRs in *N. crassa* isolates correlate with latitude as well as with circadian period. Although, their sample size is much bigger than here (143 vs. 24), they did not test for clade properties of the isolates, which might explain the differences. Villalta *et al.* (2009) described former *N. crassa* isolates from Congo as new *Neurospora* specie, which would exclude four isolates analysed by Michael and collaborators. In addition, the species belonging of two of their isolates from Pakistan and three from Thailand is not yet clear as might be also a new *Neurospora* specie, as showed here by sequencing neutral markers (Figure 3.2). Thus, Michael's *et al.* (2007) results have to be reconsidered.

Physiology

Although, the results presented here show that FRPs become longer with higher latitudes when all isolates were taken into account, this correlation disappeared when the phenotype-group 2 was excluded from the analysis. The lack of a latitudinal cline for circadian period is, also, supported by Madeti (2008), who analysed a much bigger sample size (97 isolates). The negative correlation between FRP and latitude, shown by Michael *et al.* (2007) might be due to the finding that nine of the analysed isolates potentially belong to a different *Neurospora* species or to as yet to be defined separate group (as mentioned before).

In other organisms, both positive and negative correlations between FRP and latitude were found. In *Arabidopsis thaliana* the period of the clock-mediated leave movements correlates positively with latitude (Michael *et al.*, 2003). A positive correlation was also found in the circadian rhythm of adult locomotor activity of *Drosophila ananassae* (Joshi, 1999) and in the eclosion (emergence of adults from the pupal cases) of *Drosophila auraria* (Pittendrigh and Takamura, 1989). However, a negative correlation of the pupal eclosion rhythm with latitude was found in *Drosophila littoralis* (Lakinen and Forstman, 2006).

In *Neurospora* isolates, Michael *et al.* (2007) also showed a correlation between latitude and phase of the conidiation, but in the data of this thesis no such correlation was found. This is probably due to different methods used to assess the phase. Michael *et al.* (2007) calculated the phase as the timing of the first band after the release into darkness. Thus, their phase represents the response to the release, which does not allow a determination of the phase of entrainment. In this thesis, the phase of entrainment was assessed during the light-dark and temperature cycles after it had reached the stable relationship between the onset of banding and mid-dark/cold. Similar to the present results, no correlation between the phase in the leave movements and latitude in *Arabidopsis* was found (Michael *et al.*, 2003). Still, the lack of a correlation between phase of entrainment and latitude might be a mechanism by which the *Neurospora* isolates exploit new environments and easily adapt to them.

Since *Neurospora crassa* isolates did not show a clear and well-established latitudinal cline in FRP and phase of entrainment but rather a separation into two physiological groups based on their genetic properties, *N. crassa* isolates appear not to adapt to specific environments but rather maintain the variability in their circadian properties. This may help to spread its spores fast into new environments and easily survive there. This also holds for their not well-described evolution and wide geographical distribution. The lack on a latitudinal cline in *Neurospora crassa* may also reflect that this fungus may be an opportunist in its life cycles and its propagation. Latitudinal clines and photoperiodism are generally found in organisms that adapt their biology to seasons. Although, photoperiodic responses have been shown for *Neurospora crassa* strains (e.g. *frq-1*, *frq-7*, *frq-10*;

Tan *et al.*, 2004), the general strategy of this fungus may be geared more towards responding to current environmental conditions (e.g., fires) that to an elaborate seasonal regulation.

CONCLUSIONS

In this thesis several groups of *Neurospora crassa* isolates have been identified according to their similarity in genetics (by sequencing neutral markers, clock gene *white collar-1* and the promoter of the clock gene *frequency*) and circadian phenotype (by analysing the banding pattern in different light-dark and temperature cycles). None of these groupings could be explained by their geographical distribution. Traits FRP and phase did not show a correlation with genetics but rather a separation into two phenotypic groups. Although, the circadian phenotypes could not be explained by the phylogenetic data (sequencing of neutral markers), nor by the genetic analysis of *wc-1* and *frq* promoter, the lack of a latitudinal cline both in FRP and in phase of entrainment suggests that *N. crassa* isolates do not show a mechanism for environmental adaptation.

SUMMARY

The astronomical interaction between earth, sun and moon create four temporal structures: the tides (12.5 h), the day (24 h), the lunar cycle (28.5 d) and the year (365.25 d). These are represented internally by biological clocks that enable organisms to increase their fitness by anticipating the regular changes within these environmental rhythms. The 24-h day is represented by the circadian clock, which controls physiology, metabolism and behaviour at all levels of the organism. Since circadian clocks produce internal days that are often longer or shorter than 24 h, they actively synchronize to the light-dark cycle, which itself depends both on time of year (short winter days vs long summer days) and on latitude. As biological clocks are related to fitness, they are widely used to study geographical adaptation.

The ascomycete *Neurospora crassa* produces asexual spores under circadian control, and its clock shares many similarities that of invertebrates and vertebrates. Therefore, *N. crassa* became a valuable model system for studying the molecular mechanisms of the circadian clock by studying mutants. To study the impact of environmental factors on clock evolution, different wild-type isolates are of great value. The aim of this thesis was to identify potential genetic and physiological differences in the circadian system according to the geographical distribution of wild *N. crassa* isolates.

Wild *Neurospora* isolates have been collected from all over the world. Yet, the phylogenetic relationship of some isolates is still unknown. Their relationship is, however, important to understand genetic and physiological results. The first part of this thesis therefore aimed to clarify the phylogenetic relationships. The results showed that most European isolates belonged to clade B, and Caribbean isolates to clade A. Isolates from Thailand and Pakistan formed a separate group, but its exact phylogenetic relationship still has to be determined. The Italian isolate formed an exception (belonging to clade A) and suggested human influence on the distribution of *N. crassa* isolates.

The genetic differences between isolates were studied for candidate regions, the *white collar-1 (wc-1)* gene and *frequency (frq)* promoter. *White collar-1* and *frequency* are central components of the *Neurospora* circadian feedback loop. The phylogenetic groups associated with these two candidate regions neither corresponded to clade classification nor to the geographical distribution of the isolates. Thus, different selection pressures may have acted on neutral markers (used for the clade classification) and clock genes.

Physiological differences of the circadian behaviour (asexual spore formation) were studied extensively under different entraining conditions as special adaptations may occur in relation to the natural photo- and thermo-periods. No correlation between free-running period and phase of entrainment with latitude was found, suggesting that period and phase are not traits selected by the environment. Furthermore, the circadian phenotypes of all wild *N. crassa* isolates were similar, irrespective of their phylogenetic relationship or their geographical origin, with the remarkable exception of four European isolates. This raises the intriguing question of how these phylogenetically and geographically distinct phenotypes, which are very similar in their circadian behaviour, may have evolved.

ZUSAMMENFASSUNG

Die astronomische Interaktion zwischen Erde, Sonne und Mond erzeugt vier zeitliche Strukturen: Die Gezeiten (12,5 Stunden), den Tag (24 Stunden), den Mondzyklus (28,5 Tage) und das Jahr (365,25 Tage). Sie werden intern durch biologische Uhren repräsentiert, die es dem Organismus ermöglichen, seine Fitness zu erhöhen, indem regelmäßige Veränderungen innerhalb der Umwelt-Rhythmen antizipiert werden können. Der 24 Stunden-Tag wird durch die zirkadiane Uhr repräsentiert, welche Physiologie, Metabolismus und Verhalten auf allen Ebenen des Organismus kontrolliert. Nachdem zirkadiane Uhren interne Tage erzeugen, die oft länger oder kürzer als 24 Stunden sind, müssen sie aktiv an den Licht-Dunkel-Zyklus angepasst werden, welches an sich von der Jahreszeit (kurze Wintertage vs. lange Sommertage) und dem Breitengrad abhängt. Weil biologische Uhren einen Bezug zu Fitness haben, werden sie vornehmlich auch zu Untersuchungen geographischer Adaption verwendet.

Der Ascomycet *Neurospora crassa* produziert asexuelle Sporen unter zirkadianer Kontrolle, und seine Uhr hat viele Ähnlichkeiten mit der von Wirbellosen und Wirbeltieren. So wurde *N. crassa* durch das Studium von Mutanten ein wertvolles Modellsystem für die Erforschung der molekularen Mechanismen der zirkadianen Uhr. Um den Einfluss von Umweltfaktoren auf die Uhr-Evolution zu studieren, sind verschiedene Wildtyp-Stämme von großem Wert. Das Ziel dieser Arbeit war es, mögliche genetische und physiologische Unterschiede im zirkadianen System nach der geographischen Herkunft der wilden *N. crassa* -Stämme zu identifizieren.

Wilde *Neurospora*-Stämme wurden in der ganzen Welt gesammelt. Dennoch ist die phylogenetische Verwandtschaft einiger Stämme noch unbekannt. Ihre Beziehung ist jedoch wichtig, um genetische und physiologische Ergebnisse zu verstehen. Der erste Teil dieser Arbeit hatte das Ziel, die Verwandtschaftsverhältnisse zu klären. Die Ergebnisse zeigten dass die meisten europäischen Stämmen zu Clade B und die Karibik-Stämme zu Clade A gehörten. Die Stämme aus Thailand und Pakistan bildeten eine eigene Gruppe, jedoch muss ihre genaue phylogenetische Verwandtschaft noch ermittelt werden. Da der italienische Stamm eine Ausnahme war (Clade A), weist er auf einen menschlichen Einfluss auf die Verteilung von *N. crassa* -Stämmen hin.

Die genetischen Unterschiede zwischen den Stämmen wurden an Kandidaten-Regionen des *white collar-1* (*wc-1*)-Gens und des *frequency* (*frq*)-Promotors untersucht. *White collar-1* und *frequency* sind zentrale Komponenten des zirkadianen Rückkopplungsmechanismus von *Neurospora*. Die phylogenetischen Gruppen von diesen beiden Kandidaten Regionen konnten nicht assoziiert werden, weder mit der Clade-Klassifizierung noch mit der geographischen Verteilung der Stämme. So könnten jeweils unterschiedliche Selektionsdrücke auf die neutralen Marker (verwendet für die Klassifizierung von Clades) beziehungsweise auf die Uhr-Gene gewirkt haben.

Physiologische Unterschiede im zirkadianen Verhalten (asexuelle Sporenbildung) wurden unter verschiedenen Entrainment-Bedingungen untersucht, da spezifische Anpassungen in Bezug auf natürliche Licht-und Temperatur-Perioden auftreten können. Keine Korrelation konnte zwischen der freilaufenden Periode (FRP) bzw. der Phase des Entrainment und der geographischen Breite gefunden werden, was darauf hindeutet, dass die Merkmale FRP und Phase nicht von Umweltbedingungen selektiert werden. Darüber hinaus waren die zirkadianen Phänotypen aller wilden *N. crassa*-Stämme unabhängig von ihrer phylogenetischen Verwandtschaft oder ihrer geografischen Herkunft, ähnlich, mit der bemerkenswerten Ausnahme von vier europäischen Stämme. Dies wirft die interessante Frage auf, wie diese phylogenetisch und geographisch unterschiedlichen Phänotypen, die sehr ähnlich in ihrem zirkadianen Verhalten sind, sich entwickelt haben könnten.

REFERENCES

Akerstedt, T. (**2003**) Shift work and disturbed sleep/wakefulness. *Occup. Med. (Lond.)*, 53 (2): 89-94.

Allada, R., White, N. E., So, W. V., Hall, J. C., and Rosbash, M. (**1998**) A mutant *Drosophila* homolog of mammalian *Clock* disrupts circadian rhythms and transcription of *period* and *timeless*. *Cell*, 93 (5): 791-804.

Anders, T. F. (**1982**) Biological Rhythms in Development. *Psychosomatic Medicine*, 44 (1): 61-72.

Aronson, B. D., Johnson, K. A., and Dunlap, J. C. (**1994**) Circadian clock locus *frequency*: Protein encoded by a single open reading frame defines period length and temperature compensation. *PNAS*, 91: 7683-7687.

Aronson, B. D., Lindgren, K. M., Dunlap, J. C., and Loros, J. J. (**1994a**) An efficient method for gene disruption in *Neurospora crassa*. *Mol. Gen. Genet.*, 242: 490-494.

Aschoff, J. (**1960**) Exogenous and endogenous components in circadian rhythms. *Cold Spring Harb. Symp. Quant. Biol.*, 25: 11–28.

Aschoff, J. (**1982**) The circadian rhythm of body temperature as a function of body size. In: Taylor, R., Johanson, K., Bolis, L. (eds) A comparison for animal physiology. *Cambridge University Press, Cambridge*, 173–189.

Aschoff, J. and Wever, R. (**1962**) Über Phasenbeziehungen zwischen biologischer Tagesperiodik und Zeitgeberperiodik. *Z vergl Physiol*, 46: 115-128.

Aschoff, J. and Wever, R. (**1980**) Über Reproduzierbarkeit circadianer Rhythmen beim Menschen. *Klin Wocheschr.*, 58: 323-335.

Baldauf, S. L. and Palmer, J. D. (**1993**) Animals and fungi are each other's closest relatives: congruent evidence from multiple proteins. *PNAS*, 90 (24): 11558-62.

Bell-Pedersen, D., Cassone, V. M., Earnest, D. J., Golden, S. S., Hardin, P. E., Thomas, T. L., and Zoran, M. J. (**2005**) Circadian rhythms from multiple oscillators: lessons from divers organisms. *Nature Reviews Genetics*, 6: 544-556.

Bell-Pedersen, D., Shinohara, M. L., Loros, J. J., and Dunlap, J. C. (**1996**) Circadian clock-controlled genes isolated from *Neurospora crassa* are late night- to early morning-specific. *PNAS*, 93: 13096–13101.

Binkley, S. (**1997**) Biological clocks: Your owner's manual. *Amsterdam: Overseas Publishers Association*.

Binkley, S. and **Mosher**, K. (**1985**) Direct and circadian control of sparrow behavior by light and dark. *Physiology & Behavior*, 35 (5): 785-797.

Binkley, S., Riebman, J. B. and Reilly, K. B. (**1977**) Timekeeping by the pineal gland. *Science*, 197 (4309): 1181-3.

Borkovich, K. A., Alex, L. A., Yarden, O., Freitag, M., Turner, G. E., Read, N. D., Seiler, S., Bell-Pedersen, D., Paietta, J., Plesofsky, N., Plamann, M., Goodrich-Tanrikulu, M., Schulte, U., Mannhaupt, G., Nargang, F. E., Radford, A., Selitrennikoff, C., Galagan, J. E., Dunlap, J. C., Loros, J. J., Catcheside, D., Inoue, H., Aramayo, R., Polymenis, M., Selker, E. U., Sachs, M. S., Marzluf, G. A., Paulsen, I., Davis, R., Ebbole, D. J., Zelter, A., Kalkman, E. R., O'Rourke, R., Bowring, F., Yeadon, J., Ishii, C., Suzuki, K., Sakai, W., and Pratt, R. (**2004**) Lessons from the genome sequence of *Neurospora crassa*: Tracing the path from genomic blueprint to multicellular organism. *Microbiology and Molecular Biology Reviews*, 68 (1): 1–108.

Borstnik, B. and **Pumpernik**, D (**2002**) Tandem repeats in protein coding regions of primate genes. *Genome Res.*, 12: 909-915.

Bradshaw W. E. (**1976**) Geography of photoperiodic response in diapausing mosquito. *Nature*, 262: 384-386.

Bradshaw, W. E. and **Holzapfel**, C. M. (**2007**) Evolution of Animal Photoperiodism. Annu. Rev. Ecol. Evol. Syst., 38: 1–25.

Bruns, T. (**2006**) A kingdom revised. *Nature*, 443: 758-759.

Bünning, E. (**1936**) Die endogene Tagesrhythmik als Grundlage der photoperiodischen Reaktion. *Ber. dtsch. bot. Ges.*, 54: 590–607.

Bünning, E. (**1962**) Mechanism in circadian rhythms: functional and pathological changes resulting from beats and from rhythm abnormalities. *Annals of the New York Academy of Sciences*, 98: 901–915.

Cheng, P., He, Q., He, Q., Wang, L., and Liu, Y. (**2005**) Regulation of the *Neurospora* circadian clock by an RNA helicase. *Genes and Dev.*, 19: 234-241.

Cheng, P., He, Q., Yang, Y., Wang, L., and Liu, Y. (**2003**) Functional conservation of light, oxygen, or voltage domains in light sensing. *PNAS*, 100: 5938–5943.

Cheng, P., Yang, Y., and Liu, Y. (**2001**) Interlocked feedback loops contribute to the robustness of the *Neurospora* circadian clock. *PNAS*, 98 (13): 7408–7413.

Cheng, P., Yang,Y., Gardner, K. H., and Liu, Y. (**2002**) PAS domain-mediated WC-1/WC-2 interaction is essential for maintaining the steady-state level of WC-1 and the function of both proteins in circadian clock and light responses of *Neurospora*. *Mol. Cell. Biol.*, 22 (2): 517–524.

Cheng, Y., **Gvakharia**, B., and **Hardin**, P. E. (**1998**) Two alternatively spliced transcripts from the *Drosophila period* gene rescue rhythms having different molecular and behavioral characteristics. *Mol. Cell Biol.*, 18: 6505–6514.

Costa, G. (**1996**) The impact of shift and night work on health. *Appl. Ergon.*, 27 (1): 9-16.

Covington, M. F., **Maloof**, J. N., **Straume**, M., **Kay**, S. A., and **Harmer**, S. L. (**2008**) Global transcriptome analysis reveals circadian regulation of key pathways in plant growth and development. *Genome Biol.*, 9 (8): R130.

Crosthwaite, S. K., **Dunlap**, J. C., and **Loros**, J. J. (**1997**) *Neurospora wc-1* and *wc-2*: transcription, photoresponses, and the origins of circadian rhythmicity. *Science*, 276: 763–769.

Daan, S. and **Pittendrigh**, C. S. (**1976**) A functional analysis of circadian pacemakers in nocturnal rodents. III. Heavy water and constant light: Homeostasis of frequency? *J. Comp. Physiol. A*, 106: 267-290.

Daan, S., **Beersma**, D. G., and **Borbely**, A. A. (**1984**) Timing of human sleep: recovery process gated by a circadian pacemaker. *Am. J. Physiol.*, 246 (2 Pt 2): R161-83.

Darlington, T. K., **Wager-Smith**, K., **Ceriani**, M. F., **Staknis**, D., **Gekakis**, N., **Steeves**, T. D., **Weitz**, C. J., **Takahashi**, J. S., and **Kay**, S. A. (**1998**) Closing the circadian loop: CLOCK-induced transcription of its own inhibitors *per* and *tim*. *Science*, 280 (5369): 1599-603.

Davis, R. H. (**2000**) *Neurospora*: Contributions of mode organism. *Oxford University Press*.

de Marian, J. (**1729**) Observation botanique. *Hist. Acad. Roy. Sci.*

Degli-Innocenti, F. and **Russo**, V. E. (**1984**) Isolation of new *white collar* mutants of *Neurospora crassa* and studies on their behavior in the blue-light-induced formation of protoperithecia. *J. Bacteriol.*, 159: 757–761.

Denault, D. L., **Loros**, J. J., and **Dunlap**, J. C. (**2001**) WC-2 mediates WC-1-FRQ interaction within the PAS protein-linked circadian feedback loop of *Neurospora*. *The EMBO Journal*, 20 (1 & 2): 109-117.

Denlinger, D. L. (**1986**) Dormancy in tropical insects. *Annu. Rev. Entomol.*, 31: 239-264.

Dettman, J. R., **Jacobson**, D. J., and **Taylor**, J. W. (**2003**) A multilocus genealogical approach to phylogenetic species recognition in the model *Eukaryote Neurospora*. *Evolution*, 57 (12): 2703–2720.

Dettman, J. R., Jacobson, D. J., Turner, E., Pringle, A., and Taylor, J. W. (**2003a**) Reproductive isolation and phylogenetic divergence in *Neurospora*: Comparing methods of species recognition in a model Eukaryote. *Evolution*, 57 (12): 2721–2741.

Dibner, C., Schibler, U., and Albrecht, U. (**2010**) The mammalian circadian timing system: organization and coordination of central and peripheral clocks. *Ann. Rev. Physiol.*, 72, 517–549.

Diernfellner, A. C. R., Schafmeier, T., Merrow, M. W., and Brunner, M. (**2005**) Molecular mechanism of temperature sensing by the circadian clock of *Neurospora crassa*. *Genes and Dev.*, 19: 1968-1973.

Dodd, A. N., Salathia, N., Hall, A., Kevei, E., Toth, R., Nagy, F., Hibberd, J. M., Millar, A. J., and Webb, A. A. (**2005**) Plant circadian clocks increase photosynthesis, growth, survival, and competitive advantage. *Science*, 309: 630–633.

Dong, G., Kim, Y. I. and Golden, S. S. (**2010**) Simplicity and complexity in the cyanobacterial circadian clock mechanism. *Curr. Opin. Genet. Dev.*, 20 (6): 619-25.

Dong, W., Tang, X., Yu, Y., Nilsen, R., Kim, R., Griffith, J., Arnold, J., and Schüttler, H. B. (**2008**) Systems biology of the clock in *Neurospora crassa*. *PLoS One.*, 3 (8): e3105.

Dragovic, Z., Tan, Y., Görl, M., Roenneberg, T., and Merrow, M. (**2002**) Light reception and circadian behavior in `blind' and `clock-less' mutants of *Neurospora crassa*. *EMBO J.*, 21 (14): 3643–3651.

Dunlap, J. C. (**1999**) Molecular bases for circadian clocks. *Cell*, 96: 271–290.

Dunlap, J. C. (**2006**) Proteins in the *Neurospora* circadian clockworks. *J. Biol. Chem.*, 281 (39): 28489–28493.

Dunlap, J. C. and **Loros**, J. J. (**2004**) The *Neurospora* circadian system. *J. Biol. Rhythms*, 19 (5): 414-424.

Eckel-Mahan, K. and **Sassone-Corsi**, P. (**2009**) Metabolism control by the circadian clock and *vice versa*. *Nat. Struct. Mol. Biol.*, 16: 462–467.

Edery, I. (**2000**) Circadian rhythms in a Nutshell. *Physiol. Genomics*, 3: 59-74.

Ellison, C. E., Hall, C., Kowbel, D., Welch, J., Brem, R. B., Glass, N. L., and Taylor, J. W. (**2011**) Population genomics and local adaptation in wild isolates of a model microbial eukaryote. *PNAS*, 108 (7): 2831–2836.

Espinasa, L. W. R. J. (**2006**) Conservation of retinal circadian rhythms during cavefish eye degeneration. *Evol. Dev.*, 8: 16–22.

Feldman, J. F. and Hoyle, M. N. (**1973**) Isolation of circadian clock mutants of *Neurospora crassa*. *Genetics*, 75: 605-613.

Fenn, M. G. P. and MacDonald, D. W. (**1995**) Use of middens by red foxes: risk reverses rhythms of rats. *J. Mammal.*, 76: 130–136.

Foster, R. G. and Roenneberg, T. (**2008**) Human responses to the geophysical daily, annual and lunar cycles. *Current Biology*, 18: R784–R794.

Franke, H.-D. (**1985**) On a clocklike mechanism timing lunar-rhythmic reproduction in *Typosyllis prolifera* (Polychaeta). *J. of Comparative Physiology A: Neuroethology, Sensory, Neural and Behavioral Physiology*, 156 (4): 553-561.

Franklin, K. A., Larner, V. S., and Whitelam, G. C. (**2005**) The signal transducing photoreceptors of plants. *Int. J. Dev. Biol.*, 49 (5-6): 653-64.

Froehlich, A. C., Liu, Y., Loros, J. J., and. Dunlap, J. C. (**2002**) *White collar–1*, a circadian blue light photoreceptor, binding to the *frequency* promoter. *Science*, 297: 815.

Froehlich, A. C., Loros, J. J., and Dunlap, J. C. (**2003**) Rhythmic binding of a WHITE COLLAR-containing complex to the *frequency* promoter is inhibited by FREQUENCY. *PNAS*, 100 (10): 5914–5919.

Galagan, J. E., Calvo, S. E., Borkovich, K. A., Selker, E. U., Read, N. D., Jaffe, D., FitzHugh, W., Ma, L. J., Smirnov, S., Purcell, S., Rehman, B., Elkins, T., Engels, R., Wang, S., Nielsen, C. B., Butler, J., Endrizzi, M., Qui, D., Ianakiev, P., Bell- Pedersen, D., Nelson, M. A., Werner-Washburne, M., Selitrennikoff, C. P., Kinsey, J. A., Braun, E. L., Zelter, A., Schulte, U., Kothe, G. O., Jedd, G., Mewes, W., Staben, C., Marcotte, E., Greenberg, D., Roy, A., Foley, K., Naylor, J., Stange-Thomann, N., Barrett, R., Gnerre, S., Kamal, M., Kamvysselis, M., Mauceli, E., Bielke, C., Rudd, S., Frishman, D., Krystofova, S., Rasmussen, C., Metzenberg, R. L., Perkins, D. D., Kroken, S., Cogoni, C., Macino, G., Catcheside, D., Li, W., Pratt, R. J., Osmani, S. A., DeSouza, C. P., Glass, L., Orbach, M. J., Berglund, J. A., Voelker, R., Yarden, O., Plamann, M., Seiler, S., Dunlap, J., Radford, A., Aramayo, R., Natvig, D. O., Alex, L. A., Mannhaupt, G., Ebbole, D. J., Freitag, M., Paulsen, I., Sachs, M. S., Lander, E. S., Nusbaum, C., and Birren, B. (**2003**) The genome sequence of the filamentous fungus *Neurospora crassa*. *Nature*, 422 (6934): 859-68.

Garceau, N. Y., Liu, Y., Loros, J. J., and Dunlap, J. C. (**1997**) Alternative initiation of translation and time-specific phosphorylation yield multiple forms of the essential clock protein

FREQUENCY. *Cell*, 89: 469–476.

Garner, W. W. and **Allard**, H. A. (**1922**) Photoperiodism, the response of the plant to relative length of day and night. *Science*, 55 (1431): 582-3.

Green, C. B., Takahashi, J. S., and Bass, J. (**2008**) The meter of metabolism. *Cell*, 134: 728–742.

Gwinner, E. (**1977**) Circannual rhythms in bird migration. *Ann. Rev. Ecol. Syst.*. 8: 381-405.

Harding, R. W. and **Turner**, R. V. (**1981**) Photoregulation of the carotenoid biosynthetic pathway in *albino* and *white collar* mutants of *Neurospora crassa*. *Plant Physiol.*, 68: 745-749.

Harding, R.W. and **Melles**, S. (**1984**) Genetic analysis of the phototropism of *Neurospora crassa* perithecial beaks using *white collar* and *albino* mutants. *Plant Physiol.*, 72: 996–1000.

Hastings, J. W. and **Dunlap**, J. C. (**1986**) Cell-free components in dinoflagellate bioluminescence: The paniculate activity: oscintillons; the soluble components: luciferase, luciferin, and luciferin binding protein. *Meth. Enzym.*, 133: 307-323.

Hastings, J. W. and **Sweeney**, B. M. (**1958**) A persistent diurnal rhythm of luminescence in *Gonyaulax polyedra*. *Biol. Bull.*, 115: 440-458.

Hastings, M. H., Maywood, E. S., and Reddy A. B. (**2008**) two decades of circadian time. *J. Neuroendocronolgy*, 20: 812-819.

Hastings, M. H., Reddy, A. B., and Maywood E. S. (**2003**) A clockwork web: circadian timing in brain and periphery, in health and disease. *Nat. Rev. Neurosci.*, 4: 649–661.

Hayes, D. K., Sullivan, W. N., Oliver, M. Z., and Schechter, M S. (**1970**) Photoperiod manipulation of insect diapause: a method of pest control? *Science*, 169 (943): 382-3.

He, Q. and **Liu**, Y. (**2005**) Degradation of the *Neurospora* circadian clock protein FREQUENCY through the ubiquitin-proteasome pathway. *Biochem. Soc. Trans.*, 33: 953–956.

Hong, C. I., Ruoff, P., Loros, J. J., and Dunlap, J. C. (**2008**) Closing the circadian negative feedback loop: FRQ-dependent clearance of WC-1 from the nucleus. *Genes Dev.*. 22: 3196-3204.

Huang, W., Ramsey, K. M., Marcheva, B., and Bass, J. (**2011**) Circadian rhythms, sleep, and metabolism. *J. Clin. Invest.*, 121 (6): 2133-41.

Huelsenbeck, J. P., Ronquist, F., Nielsen, R., and Bollback, J. P. (**2001**) Bayesian inference of phylogeny and its impact on evolutionary biology. *Science*, 294: 2310–2314.

Ingold, C. T. (**1971**) Periodicity in fungal spores. *Oxford University Press*, London: 214-238.

Iwasaki, H. and **Dunlap**, J. C. (**2000**) Microbial circadian oscillatory systems in *Neurospora* and *Synechococcus*. *Curr. Opin. Microbiol.*, 3: 189-196.

Jackson, R. J., Hellen, C. U., and Pestova, T. V. (**2010**) The mechanism of eukaryotic translation initiation and principles of its regulation. *Nat. Rev. Mol. Cell Biol.*, 11: 113–127.

Jacobson, D. J., Dettman, J. R., Adams, R. I., Boesl, C., Sultana, S., Roenneberg, T., Merrow, M., Duarte, M., Marques, I., Ushakova, A., Carneiro, P., Videira, A., Navarro-Sampedro, L., Olmedo, M., Corrochano, L. M., and Taylor, J. W. (**2006**) New findings of *Neurospora* in Europe and comparisons of diversity in temperate climates on continental scales. *Mycologia*, 98 (4): 550-9.

Jacobson, D. J., Powell, A. J., Dettman, J. R., Saenz, G. S., Barton, M. M., Hiltz, M. D., Dvorachek, W. H., Glass, N. L., Taylor, J. W., and Natvig, D. O. (**2004**) *Neurospora* in temperate forests of western North America. *Mycologia*, 96 (1): 66–74.

Jeffery, W. R. (**2005**) Adaptive evolution of eye degeneration in the Mexican Blind Cavefish. *J. Hered.*, 96: 185–196.

Johnsen, A., Fidler, A. E., Kuhn, S., Carter, K. L., Hoffmann, A., Barr, I. R., Biard, C., Charmantier, A., Eens, M., Korsten, P., Siitari, H., Tomiuk, J., and Kempenaers, B. (**2007**) Avian *Clock* gene polymorphism: evidence for a latitudinal cline in allele frequencies. *Mol Ecol.*, 16 (22): 4867-80.

Johnson, C. H., Elliott, J. A., and Foster, R. (**2003**) Entrainment of circadian programs. *Chronobiology International*, 20 (5): 741–774.

Joshi, D. S. (**1999**) Latitudinal variation in locomotor activity rhythm in adult *Drosophila ananassae*. *Can. J. Zool. Rev.*, 77: 865–870.

Jovani, R., Blas, J., Navarro, C., and Mougeot, F. (**2010**) Feather growth bands and photoperiod. *J. Avian Biol.*, 41: 1-5.

Kauffman, A. S., Cabrera, A., and Zucker, I. (**2001**) Energy intake and fur in summer- and winter-acclimated Siberian hamsters (*Phodopus sungorus*). *Am. J. Physiol. Regul. Integr. Comp. Physiol.*, 281: R519–R527.

Kaufman, R. A. and **Broquist**, H. P. (**1977**) Biosynthesis of carnitine in *Neurospora crassa*. *J. Biol. Chem.*, 252 (21): 1437-7439.

Kay, S. A. (**1997**) As time PASses: the first mammalian clock gene. *Science*, 276 (5315): 1093.

King, D., Zhao, Y., Sangoram, A., Wilsbacher, L., Tanaka, M., Antoch, M., Steeves, T., Vitaterna, M., Kornhauser, J., Lowrey, P., Turek, F. W., and Takahashi, J. S. (**1997**) Positional cloning of the mouse circadian clock gene. *Cell*, 89 (4): 641-53.

Kivela, A., **Kauppila**, A., Ylostalo, P., Vakkuri, O., and Leppaluoto, J. (**1988**) Seasonal, menstrual and circadian secretions of melatonin, gonadotropins and prolactin in women. *Acta. Physiol. Scand.*, 132: 321-327.

Koilraj, A. J., Sharma, V. K., Marimuthu, G., and Chandrashekaran, M. K. (**2000**) Presence of circadian rhythms in the locomotor activity of a cave-dwelling millipede *Glyphiulus cavernicolus* sulu (*Cambalidae, Spirostreptida*). *Chronobiol. Int.*, 17: 757–765.

Konopka, R. J. and **Benzer**, S. (**1971**) Clock mutants of *Drosophila melanogaster*. *PNAS*, 68 (9): 2112-6.

Kostal, V. (**2006**) Eco-physiological phases of insect diapause. *J. Insect Physiol.*, 52 (2): 113-27.

Kyriacou, C. P., Peixoto, A. A., Sandrelli, F., Costa, R., and Tauber, E. (**2007**) Clines in clock genes: fine-tuning circadian rhythms to the environment. *Trends in Genetic*, 24 (3): 124-132.

Lakin-Thomas, P. L., Brody, S., and Cote, G. G. (**1991**) Amplitude model for the effects of mutations and temperature on period and phase resetting of the *Neurospora* circadian oscillator. *J. Biol. Rhythms*, 6: 281-297.

Lankinen, P. and **Forsman**, P. (**2006**) Independence of genetic geographical variation between photoperiodic diapause, circadian eclosion rhythm, and Thr-Gly repeat region of the *period* gene in *Drosophila littoralis*. *J. Biol. Rhythms*, 21: 3–12.

Lauter, F. R., Yamashiro, C. T., and Yanofsky, C. (**1997**) Light stimulation of conidiation in *Neurospora crassa*: studies with the wild-type strain and mutants *wc-1*, *wc-2* and *acon-2*. *J. Photochem. Photobiol. B*, 37 (3): 203–211.

Lee, K., Loros, J. J., and Dunlap, J. C. (**2000**) Interconnected feedback loops in the *Neurospora* circadian system. *Science*, 289: 107–110.

Lees, A. (**1973**) Photoperiodic time measurements in the aphid *Megoura Viciae*. *J. Insect Physiol.*, 19: 2279-2316.

Levine, M. E., Milliron, A. N., and Duffy, J. (**1994**) Diurnal and seasonal rhythms of melatonin, cortisol and testosterone in interior Alaska. *Arctic Med. Res.*, 53: 25-34.

Levinson, G. and **Gutman**, G. A. (**1987**) Slipped-Strand mispairing: A major mechanism for DNA sequence evolution. *Mol. Biol. Evol.*, 4 (3): 203-221.

Lewis, M. T. and **Feldman**, J. F. (**1996**) Evolution of the *frequency* (*frq*) clock locus in Ascomycete Fungi. *Mol. Biol. Evol.*, 13 (9): 1233-1241.

Liu, Y. (2003) Molecular mechanisms of entrainment in the *Neurospora* circadian clock. *J Biol. Rhythms*, 18: 195-205.

Liu, Y. and **Bell-Pedersen**, D. (2006) Circadian rhythms in *Neurospora crassa* and other filamentous fungi. *Eukaryotic cell*, 5 (8): 1184–1193.

Liu, Y., Garceau, N. Y., Loros, J. J., and Dunlap, J. C. (**1997**) Thermally regulated translational control of FRQ mediates aspects of temperature responses in the *Neurospora* circadian clock. *Cell*, 89: 477–486.

Liu, Y., Loros, J., and Dunlap, J. C. (**2000**) Phosphorylation of the *Neurospora* clock protein FREQUENCY determines its degradation rate and strongly influences the period length of the circadian clock. *PNAS*, 97: 234-239.

Loros, J. J. and **Feldman**, J. F. (**1986**) Loss of temperature compensation of circadian period length in the *frq-9* mutant of *Neurospora crassa*. *J. Biol. Rhythms*, 1 (3): 187-198.

Loros, J. J., Richman, A., and Feldman, J. F. (**1986**) A recessive circadian clock mutant at the *frq* locus of *Neurospora crassa*. *Genetics*, 114: 1095-1110.

Maddison, D. R. and **Maddison**, W. P. (**2000**) MacClade 4: Analysis of phylogeny and character evolution. Version 4.0. Sinauer Associates, Sunderland, Massachusetts.

Madeti, C. (**2008**) The clock in the cell: Entrainment of the circadian clock in *Neurospora crassa*. Dissertation.

Manjunatha, T., Hari Dass, S. and Sharma, V. K. (**2008**) Egg-laying rhythm in *Drosophila melanogaster*. *J. Genet*. 87: 495–504.

Merrow, M. and **Dunlap**, J. C. (**1994**) Intergenic complementation of a circadian rhythmicity defect: phylogenetic conservation of structure and function of the clock gene *frequency*. *EMBO J.*, 13: 2257-2266.

Merrow, M., Brunner, M., and Roenneberg, T. (**1999**) Assignment of circadian function for the *Neurospora* clock gene *frequency*. *Nature*, 399: 584-586.

Merrow, M., Spoelstra, K., and Roenneberg, T. (**2005**) The circadian cycle: daily rhythms from behavior to genes. *EMBO reports*, 6 (10): 930-935.

Michael, T. P., Park, S., Kim, T. S., Booth, J., Byer, A., Sun, Q., Chory, J., and Lee, K. (**2007**) Simple sequence repeats provide a substrate for phenotypic variation in the *Neurospora crassa* circadian clock. *PLoS One*, 2 (8): e795.

Michael, T. P., Salome, P. A., Yu, H. J., Spencer, T. R., Sharp, E. L., McPeek, M. A., Alonso, J.

M., Ecker, J. R., and McClung, C. R. (**2003**) Enhanced fitness conferred by naturally occurring variation in the circadian clock. *Science*, 302: 1049-53.

Müller, G. M. and **Schmit**, J. P. (**2007**) Fungal biodiversity: what do we know? What can we predict? *Biodiversity and Conservation*, 16 (1): 1-5.

Moser, M., Fruhwirth, M., Penter, R., and Winker, R. (**2006**) Why life oscillates – from a topographical towards a functional chronobiology. *Cancer Causes Control*, 17: 591–599.

Nakajima, M., Imai, K., Ito, H., Nishiwaki, T., Murayama, Y., Iwasaki, H., Oyama, T., and Kondo, T. (**2005**) Reconstitution of circadian oscillation of cyanobacterial KaiC phosphorylation *in vitro*. *Science*, 308 (5720): 414-5.

Naylor, E. (**1996**) Crab clockwork: the case for interactive circatidal and circadian oscillators controlling rhythmic locomotor activity of *Carcinus maenas*. *Chronobiol. Int.* 13 (3): 153-61.

Nowrousian, M., Duffield, G., Loros, J. J., and Dunlap, J. C. (**2003**) The *frequency* gene is required for temperature dependent regulation of many clock-controlled genes in *Neurospora crassa*. *Genetics*, 164: 922-933.

Nygren, K., Strandberg, R., Wallberg, A., Nabholz, B., Gustafsson, T., García, D., Cano, J., Guarro, J., and Johannesson, H. (**2011**) A comprehensive phylogeny of *Neurospora* reveals a link between reproductive mode and molecular evolution in fungi. *Mol. Phylogenetics and Evolution*, 59: 649–663.

O'Malley, K. G. and **Banks**, M. A. (**2008**) A latitudinal cline in the Chinook salmon (*Oncorhynchus tshawytscha*) *Clock* gene: evidence for selection on PolyQ length variants. *Proc. R. Soc. B*, 275: 2813–2821.

O'Neill, J. S. and **Reddy**, A. B. (**2011**) Circadian clocks in human red blood cells. *Nature*, 469 (7331): 498-503.

Oda, K. and **Hasunuma**, K. (**1997**) Genetic analysis of signal transduction through light-induced protein phosphorylation in *Neurospora crassa* perithecia. *Mol. Gen. Genet.*, 256: 593–601.

Ouyang, Y., Andersson, C. R., Kondo, T., Golden, S. S., and Johnson, C. H. (**1998**) Resonating circadian clocks enhance fitness in cyanobacteria. *PNAS*, 95: 8660–8664.

Pandit, A. and **Maheshwari**, R. (**1994**) Sexual reproduction by *Neurospora* in nature. *Fungal Genetics Newsletter*, 41: 67-68.

Paranjpe, D. A. and **Sharma**, V. K. (**2005**) Evolution of temporal order in living organisms. *J. Circa. Rhythms*, 3: 7.

Perkins, D. D. (**1973**) Freezing as a convenient method for preserving vegetative stocks. *Neurospora Newslett.*, 20: 33.

Perkins, D. D. and Turner, B. C. (**1988**) *Neurospora* from natural populations: toward the population biology of a haploid eukaryote. *Exp. Mycol.*, 12: 91-131.

Perkins, D. D., Radford, A., Newmeyer, D., and Bjorkman, M. (**1982**) Chromosomal loci of *Neurospora crassa*. *Microbiol. Rev.*, 46: 426–570.

Perkins, D. D., Turner, B. C., and Barry, E. G. (**1976**) Strains of *Neurospora* collected from nature. *Evolution*, 30 (2): 281-313.

Pittendrigh, C. S. ed. (**1981**) Circadian systems: entrainment. *Plenum Publ.Corp.*, New York.

Pittendrigh, C. S. (**1993**) Temporal organization: Reflections of a Darwinian clock –watcher. *Annu. Rev. Physiol.*, 55: 17-54.

Pittendrigh, C. S. and Minis, D. H. (**1964**) The entrainment of circadian oscillations by light and their role as photoperiodic clocks. *Amer. Naturalist*, 98: 261-294.

Pittendrigh, C. S. and Takamura, T. (**1989**) Latitudinal clines in the properties of a circadian pacemaker. *J. Biol. Rhythms*, 4: 217–235.

Pittendrigh, C. S., Bruce, V. G., Rosensweig, N. S., and Rubin, M. L. (**1959**) Growth patterns in *Neurospora crassa*. *Nature*, 184: 169-170.

Pohl, H. (**1987**) Control of annual rhythms of reproduction and hibernation by photoperiod and temperature in the Turkish hamster. *J. Thermal Bio.*, 12 (2): 119-123.

Posada, D. and Crandall, K. A. (**1998**) Modeltest: testing the model of DNA substitution. *Bioinformatics*, 14 (9): 817-818.

Powell, A. J., Jacobson, D. J., Salter, L., and Natvig, D. O. (**2003**) Variation among natural isolates of *Neurospora* on small spatial scales. *Mycologia*, 95 (5): 809–819.

Pregueiro, A. M., Price-Lloyd, N., Bell-Pedersen, D., Heintzen, C., Loros, J. J., and Dunlap, J. C. (**2005**) Assignment of an essential role for the *Neurospora frequency* gene in circadian entrainment to temperature cycles. *PNAS*, 102 (6): 2210–2215.

Rajaratnam, S. M. and Arendt, (**2001**) Health in a 24-h society. *J. Lancet.*, 358 (9286): 999-1005.

Raven, P. H., Evert, R. F., and Eichhorn, S. E. (**1999**) Biology of Plants. 6th ed. New York: W. H. Freeman and Company.

Rémi, J., Merrow, M., and Roenneberg, T. (**2010**) A circadian surface of entrainment: varying T, τ and photoperiod in *Neurospora crassa*. *J. Biol. Rhythms*, 25 (5): 318-28.

Reppert, S. M. (**2006**) A colorful model of the circadian clock. *Cell*, 124 (2): 233-6.

Ridley, M. (**2003**) Speciation - What is the role of reinforcement in speciation? Adapted from *Evolution* 3rd edition (Boston: Blackwell Science).

Roenneberg, T. and **Aschoff**, J. (**1990**) Annual rhythm of human reproduction: I. Biology, sociology, or both? *J. Biol. Rhythms*, 5 (3): 195-216.

Roenneberg, T and **Merrow**, M. (**2002**) "What watch?... such much!" Complexity and evolution of circadian clocks. *Cell Tissue Res.*, 309: 3–9.

Roenneberg, T. and **Merrow**, M. (**2003**) The network of time: understanding the molecular circadian system. *Current Biology*, 13: 198–207.

Roenneberg, T and **Merrow**, M. (**2005**) Circadian clocks — the fall and rise of physiology. *Nature Reviews: Mol. Cell. Biol.*, 6: 965-971.

Roenneberg, T. and **Taylor**, W. (**2000**) Automated recordings of bioluminescence with special reference to the analysis of circadian rhythm. *Methods Enzymo.*, 305: 104-19.

Roenneberg, T., Daan, S., and Merrow, M. (**2003**) The Art of entrainment. *J. Biol. Rhythms*, 18 (3): 183-194.

Roenneberg, T., Dragovic, Z., and Merrow, M. (**2005**) Demasking biological oscillators: Properties and principles of entrainment exemplified by the *Neurospora* circadian clock. *PNAS*, 102 (21): 7742–7747.

Roenneberg, T., Hut, R., Daan, S., and Merrow, M. (**2010**) Entrainment concepts revisited. *J. Biol. Rhythms*, 25 (5): 329-39.

Roenneberg, T., Merrow, M., and Eisensamer, B. (**1998**) Cellular mechanisms of circadian systems. *Zoology*, 100: 273-286.

Roenneberg, T., Rémi, J., and Merrow, M. (**2010a**) Modeling a circadian surface. *J. Biol. Rhythms*, 25 (5): 340-9.

Roenneberg, T., Wirz-Justice, A., and Merrow, M. (**2003**) Life between clocks – daily temporal patterns of human chronotypes. *J. Biol. Rhythms*, 18: 80-90.

Ronquist, F., and **Huelsenbeck**, J. P. (**2003**) MRBAYES 3: Bayesian phylogenetic inference under mixed models. *Bioinformatics*, 19: 1572–1574.

Rosenblum, E. B., Römpler, H., Schöneberg, T., and Hoekstra, H. E. (**2010**) Molecular and functional basis of phenotypic convergence in white lizards at White Sands. *PNAS*, 107: 2113–2117.

Rowan, W. (**1925**) Relation of light to bird migration and developmental changes. *Nature*, 115: 494-495.

Ruoff, P., Loros, J. J., and Dunlap, J. C. (**2005**) The relationship between FRQ-protein stability and temperature compensation in the *Neurospora* circadian clock. *PNAS*, 102: 17681-17686.

Rusak, B. and Zucker, I. (**1975**) Biological rhythms and animal behavior. *Annu. Rev. Psychol.*, 26: 137-171.

Rutila, J. E., Suri, V., Le, M., So, W. V., Rosbash, M., and Hall, J. C. (**1998**) CYCLE is a second bHLH-PAS clock protein essential for circadian rhythmicity and transcription of *Drosophila period* and *timeless*. *Cell*, 93 (5): 805-14.

Sargent, M. L. and Briggs, W. R. (**1967**) The effects of blue light on a circadian rhythm of conidiation in *Neurospora*. *Plant Physiol.*, 42: 1504–1510.

Sargent, M. L., Briggs, W. R., and Woodward D. O. (**1966**) Circadian nature of a rhythm expressed by an invertaseless strain of *Neurospora crassa*. *Plant Physiol.*, 41: 1343-1349.

Saunders, D. S. (**1971**) The temperature-compensated photoperiodic clock "programming" development and pupal diapause in the flesh-fly, *Sarcophaga argyrostoma*. *J. Insect Physiol.*, 17: 801-812.

Saunders, D. S. and Bertossa, R. C. (**2011**) Deciphering time measurement: The role of circadian 'clock' genes and formal experimentation in insect photoperiodism. *J. Insect Physiol.*, 57 (5): 557-66.

Sawyer, L. A., Hennessy, J. M., Peixoto, A. A., Rosato, E., Parkinson, H., Costa, R., and Kyriacou, C. P. (**1997**) Natural variation in a *Drosophila* clock gene and temperature compensation. *Science*, 278.

Schernhammer, E. S., Laden, F., Speizer, F. E., Willett, W. C., Hunter, D. J., Kawachi, I., and Colditz, G. A. (**2001**) Rotating night shifts and risk of breast cancer in women participating in the nurses' health study. *J. Natl. Cancer Inst.*, 93 (20): 1563-8.

Sharma, V. K. (**2003**) Adaptive significance of circadian clocks. *Chronobiol. Int.*, 20 (6): 901–919.

Shear, C. L. and Dodge, B. O. (**1927**) Life histories and heterothallism of the red bread-mold fungi of the *Mornilia sitophila* group. *J. Agr. Res.*, 34: 1019- 1042.

Siegel, P. V., Gerathewohl, S. J., and Mohler, S.R. (**1969**) Time-zone effects. *Science*, 164 (885): 1249-55.

Stamatakis, A. (2006) RAxML-VI-HPC: maximum likelihood-based phylogenetic analyses with thousands of taxa and mixed models. *Bioinformatics*, 22: 2688–2690.

Stamatakis, A., Hoover, P., and Rougemont, J. (2008) A rapid bootstrap algorithm for the RAxML Web servers. *Syst. Biol.*, 57: 758–771.

Swade, R. H. (1969) Circadian rhythms in fluctuating light cycles: toward a new model of entrainment. *J. Theor. Biol.*, 24: 227-239.

Sweeney, B. M. (1976) Circadian rhythms in corals, particularly *fungiidae*. *Biol. Bull.*, 151: 236-46.

Swofford, D. L. (1998) PAUP*. Phylogenetic Analysis Using Parsimony (*and Other Methods). Version 4. Sinauer Associates, Sunderland, Massachusetts.

Syed, S., Saez, L., and Young, M. W. (2011) Kinetics of doubletime kinase-dependent degradation of the *Drosophila* period protein. *J. Biol. Chem.*, 286 (31): 27654-62.

Takahashi, J. S., Hong, H. K., Ko, C. H., and McDearmon, E. L. (2008) The genetics of mammalian circadian order and disorder: implications for physiology and disease. *Nat. Rev. Genet.*, 9: 764–775.

Tamura, K., Peterson, D., Peterson, N., Stecher, G., Nei, M., and Kumar, S. (2011) MEGA 5: Molecular Evolutionary Genetics Analysis using maximum likelihood, evolutionary distance, and maximum parsimony methods. *Molec. Biol. Evol.*, 28: 2731-2739.

Tataroğlu, O. and Schafmeier, T. (2010) Of switches and hourglasses: regulation of subcellular traffic in circadian clocks by phosphorylation. *EMBO reports*, 11: 927–935.

Tauber, E., Last, K. S., Olive, P. J. W., and Kyriacou, C. P. (2004) Clock gene evolution and functional divergence. *J. Biol. Rhythms*, 19 (5): 445-458.

Tauber, E., Roe, H., Costa, R., Hennessy, J. M., and Kyriacou, C. P. (2003) Temporal mating isolation driven by a behavioral gene in *Drosophila*. *Curr. Biol.*, 13: 140–145.

Taylor, J. W., Turner, E., P. Townsend, J. P., Dettman, J. R., and Jacobson, D. (2006) Eukaryotic microbes, species recognition and the geographic limits of species: examples from the kingdom Fungi. *Phil. Trans. R. Soc. B*, 361: 1947–1963.

Thain, S. C., Murats, G., Lynn, J. R., McGrath, R. B., and Millar, A. J. (2002) The circadian clock that controls gene expression in *Arabidopsis* is tissue specific. *Plant Physiol.*, 130 (1): 102-10.

Turek, F. W. (2007) From circadian rhythms to clock genes in depression. *Int. Clin. Psychopharmacol.*, 22: S1–S8.

Turner, B. C., Perkins, D. D., and Fairfield, A. (**2001**) *Neurospora* from natural populations: A global study. *Fungal Genetics and Biology*, 32: 67–92.

Turner, E., Jacobson, D. J., and Taylor, J. W. (**2010**) Reinforced postmating reproductive isolation barriers in *Neurospora*, an *Ascomycete* microfungus. *J. Evol. Biol.*, 23: 1642–1656.

Underwood, H. and **Hall**, D. (1982) Photoperiodic control of reproduction in the male lizard *Anolis carolinensis*. *J. Comp. Physiol. A*, 146 (4): 485-492.

Van Cauter, E. and **Turek**, F. W. (**1986**) Depression: A disorder of timekeeping? *Perspect. Biol. Med.*, 29: 510–519.

Vaz Nunes, M. and **Saunders**, D. (**1999**) Photoperiodic Time Measurement in Insects: A Review of Clock Models. *J. Bio. Rhythms*, 14 (2): 84-104.

Veerman, A. and **Vaz Nunes**, M. (**1987**) Analysis of the operation of the photoperiodic counter provides evidence for hourglass time measurement in the spider mite *Tetranychus urticae*. *J. Comparative Physiology A*, 160: 421–430.

Villalta, C. F., Jacobson, D. J., and Taylor, J. W. (**2009**) Three new phylogenetic and biological *Neurospora* species: *N. hispaniola*, *N. metzenbergii* and *N. perkinsii*. *Mycologia*, 101 (6): 777–789.

Vitaterna, M. H., King, D. P., Chang, A. M., Kornhauser, J. M., Lowrey, P. L., McDonald, J. D., Dove, W. F., Pinto, L. H., Turek, F. W., and Takahashi, J. S. (**1994**) Mutagenesis and mapping of a mouse gene, *Clock*, essential for circadian behavior. *Science*, 264 (5159): 719-25.

Vogel, H. J. (**1956**) A convenient growth medium for *Neurospora* (Medium N). *Microb. Genet. Bull.*, 13: 42–43.

Whittaker, R. H. (**1969**) New concepts of kingdoms or organisms. Evolutionary relations are better represented by new classifications than by the traditional two kingdoms. *Science*, 163 (3863): 150–60.

Wittmann, M., Dinich, J., Merrow, M., and Roenneberg, T. (**2006**) Social jetlag: Misalignment of biological and social time. *Chronobiology International*, 23 (1&2): 497–509.

Woelfle, M. A., Ouyang, Y., Phanvijhitsiri, K., and Johnson, C.H. (**2004**) The adaptive value of circadian clocks: an experimental assessment in cyanobacteria. *Curr. Biol.*, 14: 1481–1486.

Yakir, E., Hilam, D., Hassidim, M. and Green, R. M. (**2007**) CIRCADIAN CLOCK ASSOCIATED1 transcript stability and the entrainment of the circadian clock in *Arabidopsis*. *Plant Physiol.*, 145 (3): 925-32.

Yerushalmi, S. and **Green**, R. M. (**2009**) Evidence for the adaptive significance of circadian rhythms. *Ecology Letters*, 12: 970–981.

Yerushalmi, S., Bodenhaimer, S., and Bloch, G. (**2006**) Developmentally determined attenuation in circadian rhythms links chronobiology to social organization in bees. *J. Exp. Biol.*, 209: 1044–1051.

Zamorzaeva, I., Rashkovetsky, E., Nevo, E., and Korol, A. (**2005**) Sequence polymorphism of candidate behavioral genes in *Drosophila melanogaster* flies from 'Evolution canyon'. *Mol. Ecol.*, 14: 3235–3245.

APPENDIX

Abbreviations

τ	tau, internal period
τ_E	entrained period
µE	micro Einstein
µl	microliter
µM	micromolar
a	arrhythmic
ATP	adenosine triphosphate
bd	band
bp	base pair
BSR	biological species recognition
ccg	clock controlled gene
CIRC	Circadian Integrated Response Characteristic
CK2	casein kinase 2
cm	centimeter
CpolyQH	carboxyl-terminal polyglutamine-histidine
CRPP	circadian rhythm of photoperiodic photosensitivity
CRY, cry	CRYPTOCHROM
DD	constant darkness
DNA	dexyribonucleic acid
dNTP	deoxynucleotide triphosphate
dpi	dots per inch
E	east
EDTA	ethylenediaminetetraacetic acid
FFC	FRQ-FRH complex
FGSC	Fungal Genetic Sock Center
FLO	FRQ-less oscillator
FRH	FRQ – interaction RNA helicase
FRP	free-running period
FRQ, frq	FREQUENCY
fwd	forward
FWD	F-box/WD-40 repeat-containing protein
FWO	*frq/wc* oscillator

g	gram
h	hour
indel	insertion/deletions
l	liter
Lat	Latin
LD	light-dark
LDP	long-day plant
LL	constant light
LOV	light, oxygen, and voltage
M	molar
m^2	square meter
mat	mating type
mg	microgram
min	minute
ML	maximum likelihood
ml	milliliter
mm	millimeter
mM	micromolar
mRNA	message RNA
MSR	morphological species recognition
N	north
NpolyQ	amino-terminal polyglutamine
OD	optical density
PAS	period, aryl hydorcabon nuclear tanslocator, single-minded
PCR	polymerase chain reaction
PER, per	PERIOD
Phi on	onset of conidiation
pmol	picomolar
PRC	phase response curve
PSR	phylogenetic species recognition
rev	reverse
RNA	ribonucleic acid
rpm	rotations per minute
S	south
SCN	suprachiasmatic nucleus

SDP	short-day plant
SDS	sodium dodecyl sulfate
sec	second
seq	sequencing
SNP	single nucleotide polymorphism
SSR	simple sequence repeat
T	cycle length
Taq	Thermus aquaticus
Thr-Gly	threonine–glycine
U	unit
VRC	velocity response curve
W	watt
W	west
WC, wc	white collar
WCC	white collar complex
X	per

Recipes

Minimal Medium

20 ml	Vogels' salt (50X)
5 g	arginine
10 ng/ml	biotin
20 g	glucose
500 ml	H_2O

Race tube media

2%	Bacto-Agar
0.5%	arginine
1X	Vogels' salts
10 ng/ml	biotin

Trace elements solution

0.05 g/l	H_3BO_3 (anhydrous)
0.25 g/l	$CuSO_4$ x $5H_2O$
1.0 g/l	$Fe(NH_4)_2(SO_4)$ x $6H_2O$
0.05 g/l	$MnSO_4$ x $5H_2O$
5.0 g/l	$ZnSO_4$ x $7H_2O$
0.05 g/l	Na_2MoO_4 x $2H_2O$
5.0 g/l	citric acid x $1H_2O$

50 X Vogel's salts

150 g	Na_3citrate x $5.5H_2O$ ($2H_2O$)
250 g	KH_2PO_4 (anhydrous)
100 g	NH_4NO_3 (anhydrous)
10 g	$MgSO_4$ x $7H_2O$
5 g	$CaCl_2$ x $2H_2O$ ($0H_2O$)
5.0 ml	Trace elements solution
2.5 ml	biotin (0.1 mg/ml)

Total volume 1000 ml. In 750 ml H_2O dissolve successively. Make sure ingredients dissolved completely before adding the next one. Adjust pH to 5.8. Add 2 ml chloroform

for preservation | 2% | glucose
| 1X | Vogel's salts
Vogel's minimal medium | 2% | agar

i want morebooks!

Buy your books fast and straightforward online - at one of world's fastest growing online book stores! Environmentally sound due to Print-on-Demand technologies.

Buy your books online at
www.get-morebooks.com

Kaufen Sie Ihre Bücher schnell und unkompliziert online – auf einer der am schnellsten wachsenden Buchhandelsplattformen weltweit! Dank Print-On-Demand umwelt- und ressourcenschonend produziert.

Bücher schneller online kaufen
www.morebooks.de

VDM Verlagsservicegesellschaft mbH
Heinrich-Böcking-Str. 6-8   Telefon: +49 681 3720 174   info@vdm-vsg.de
D - 66121 Saarbrücken   Telefax: +49 681 3720 1749   www.vdm-vsg.de

Printed by Books on Demand GmbH, Norderstedt / Germany